# 80 Advances in Polymer Science

# Epoxy Resins and Composites IV

Editor: K. Dušek

With Contributions by
R. Dave, M. P. Duduković, H. Furukawa,
T. Kamon, J. L. Kardos, E. S.-W. Kong,
E. F. Oleinik, S. D. Senturia, N. F. Sheppard, Jr.

With 145 Figures and 19 Tables

Springer-Verlag  Berlin  Heidelberg  New York
London  Paris  Tokyo

ISBN-3-540-16423-5 Springer-Verlag Berlin Heidelberg New York
ISBN-0-387-16423-5 Springer-Verlag New York Heidelberg Berlin

Library of Congress Catalog Card Number 61-642

This work is subject to copyright. All rights are reserved, whether the whole or part of the material is concerned, specifically those of translation, reprinting, re-use of illustrations, broadcasting, reproduction by photocopying machine or similar means, and storage in data banks. Under § 54 of the German Copyright Law where copies are made for other than private use, a fee is payable to "Verwertungsgesellschaft Wort". Munich.

© Springer-Verlag Berlin Heidelberg 1986
Printed in GDR

The use of registered names, trademarks, etc. in this publication does not imply, even in the absence of a specific statement, that such names are exempt from the relevant protective laws and regulations and therefore free for general use.

Typesetting and Offsetprinting: Th. Müntzer, GDR;
Bookbinding: Lüderitz & Bauer, Berlin

2154/3020-543210

# Editors

Prof. Henri Benoit, CNRS, Centre de Recherches sur les Macromolecules, 6, rue Boussingault, 67083 Strasbourg Cedex, France

Prof. Hans-Joachim Cantow, Institut für Makromolekulare Chemie der Universität, Stefan-Meier-Str. 31, 7800 Freiburg i. Br., FRG

Prof. Gino Dall'Asta, Via Pusiano 30, 20137 Milano, Italy

Prof. Karel Dušek, Institute of Macromolecular Chemistry, Czechoslovak Academy of Sciences, 16206 Prague 616, ČSSR

Prof. Hiroshi Fujita, Department of Macromolecular Science, Osaka University, Toyonaka, Osaka, Japan

Prof. Manfred Gordon, Department of Pure Mathematics and Mathematical Statistics, University of Cambridge CB2 1SB, England

Prof. Gisela Henrici-Olivé, Chemical Department, University of California, San Diego, La Jolla, CA 92037, U.S.A.

Prof. Dr. habil. Günter Heublein, Sektion Chemie, Friedrich-Schiller-Universität, Humboldtstraße 10, 69 Jena, DDR

Prof. Dr. Hartwig Höcker, Deutsches Wollforschungs-Institut e. V. an der Technischen Hochschule Aachen, Veltmanplatz 8, 5100 Aachen, FRG

Prof. Hans-Henning Kausch, Laboratoire de Polymères, Ecole Polytechnique Fédérale de Lausanne, 32, ch. de Bellerive, 1007 Lausanne, Switzerland

Prof. Joseph P. Kennedy, Institute of Polymer Science, The University of Akron, Akron, Ohio 44325, U.S.A.

Prof. Anthony Ledwith, Department of Inorganic, Physical and Industrial Chemistry, University of Liverpool, Liverpool L69 3BX, England

Prof. Seizo Okamura, No. 24, Minamigoshi-Machi Okazaki, Sakyo-Ku. Kyoto 606, Japan

Professor Salvador Olivé, Chemical Department, University of California, San Diego, La Jolla, CA 92037, U.S.A.

Prof. Charles G. Overberger, Department of Chemistry. The University of Michigan, Ann Arbor, Michigan 48 104, U.S.A.

Prof. Helmut Ringsdorf, Institut für Organische Chemie, Johannes-Gutenberg-Universität, J.-J.-Becher Weg 18–20, 6500 Mainz, FRG

Prof. Takeo Saegusa, Department of Synthetic Chemistry, Faculty of Engineering, Kyoto University, Kyoto, Japan

Prof. John L. Schrag, University of Wisconsin, Department of Chemistry, 1101 University Avenue, Madison, Wisconsin 53706, U.S.A.

Prof. Günter Victor Schulz, Institut für Physikalische Chemie der Universität, 6500 Mainz, FRG

Editors

Prof. William P. Slichter, Chemical Physics Research Department, Bell Telephone Laboratories, Murray Hill, New Jersey 07971, U.S.A.

Prof. John K. Stille, Department of Chemistry. Colorado State University, Fort Collins, Colorado 80523, U.S.A.

# Editorial

With the publication of Vol. 51 the editors and the publisher would like to take this opportunity to thank authors and readers for their collaboration and their efforts to meet the scientific requirements of this series. We appreciate the concern of our authors for the progress of "Advances in Polymer Science" and we also welcome the advice and critical comments of our readers.

With the publication of Vol. 51 we would also like to refer to a editorial policy: *this series publishes invited, critical review articles of new developments in all areas of polymer science in English (authors may naturally also include workes of their own)*. The responsible editor, that means the editor who has invited the author, discusses the scope of the review with the author on the basis of a tentative outline which the author is asked to provide. The author and editor are responsible for the scientific quality of the contribution.

Manuscripts must be submitted in content, language and form satisfactory to Springer-Verlag. Figures and formulas should be reproducible. To meet the convenience of our readers, the publisher will include a "volume index" which characterizes the content of the volume.

The editors and the publisher will make all efforts to publish the manuscripts as rapidly as possible. Contributions from diverse areas of polymer science must occasionally be united in one volume. In such cases a "volume index" cannot meet all expectations, but will nevertheless provide more information than a mere volume number.

Starting with Vol. 51, each volume will contain a subject index.

Editors                                                                                  Publisher

# Preface

This volume 80 of ADVANCES IN POLYMER SCIENCE contains the fourth part of a series of critical reviews on selected topics concerning epoxy resins and composites. The last decade has been marked by an intense development of applications of epoxy resins in traditional and newly developing areas such as coatings, adhesives, civil engineering or electronics and high-performance composites. The growing interest in applications and requirements of high quality and performance has provoked a new wave in fundamental research in the area of resin synthesis, curing systems, properties of cured products and methods of their characterization.

The collection of reviews to be published in ADVANCES IN POLYMER SCIENCE is devoted just to these fundamental problems. The epoxy resin-curing agent formulations are typical thermosetting systems of a rather high degree of complexity. Therefore, some of the formation-structure-properties relationships are still of empirical or semiempirical nature. The main objective of this series of articles is to demonstrate the progress in research towards the understanding of these relationships in terms of current theories of macromolecular systems.

Because of the complexity of the problems discussed, the theoretical approaches and interpretation of results presented by various authors and schools may be somewhat different. It may be hoped, however, that a confrontation of ideas may positively contribute to the knowledge about this important class of polymeric materials.

In view of the wide range of this volume, it was not possible to publish all contributions in successive volumes of ADVANCES IN POLYMER SCIENCE. Part I of the articles is published in Vol. 72; Part II appeared in Vol. 75 and Part III in Vol. 78.

The reader may appreciate receiving a list of all contributions to EPOXY RESINS AND COMPOSITES I–IV appearing in ADVANCES IN POLYMER SCIENCE:

M. T. Aronhime and J. K. Gillham:

A. Apicella and L. Nicolaïs (University of Naples, Naples, Italy)
The Time-Temperature-Transformation (TTT) Cure Diagram of Thermosetting Polymeric Systems.
Effect of Water on the Properties of Epoxy Matrix and Composites (Part I, Vol. 72).

J. M. Barton (Royal Aircraft Establishment, Farnborough, UK):
The Application of Differential Scanning Calorimetry (DSC) to the Study of Epoxy Resins Curing Reactions (Part I, Vol. 72).

L. T. Drzal (Michigan State University, East Lansing, MI, USA)
The Interphase in Epoxy Composites (Part II, Vol. 75).

K. Dušek (Institute of Macromolecular Chemistry, Czechoslovak Academy of Sciences, Prague, Czechoslovakia).
Network Formation in Curing of Epoxy Resins (Part III, Vol. 78).

T. Kamon and H. Furukawa (The Kyoto Municipal Research Institute of Industry, Kyoto, Japan).
Curing Mechanism and Mechanical Properties of Cured Epoxy Resins (Part IV, Vol. 80).

J. L. Kardos and M. P. Duduković (Washington University, St. Louis. MO, USA).
Void Growth and Transport During Processing of Thermosetting Matrix Composites (Part IV, Vol. 80).

A. J. Kinloch (Imperial College, London, UK).
Mechanics and Mechanisms of Fracture of Thermosetting Epoxy Polymers (Part I, Vol. 72).

E. S. W. Kong (Hewlett-Packard Laboratories, Palo Alto, CA, USA).
Physical Aging in Epoxy Matrices and Composites (Part IV, Vol. 80).

J. D. LeMay and F. N. Kelley (University of Akron, Akron, OH, USA).
Structure and Ultimate Properties of Epoxy Resins (Part III, Vol. 78).

F. Lohse, and H. Zweifel (Ciba-Geigy, Basle, Switzerland).
Photocrosslinking of Epoxy Resins (Part III, Vol. 78).

E. Mertzel and J. L. Koenig (Case Western Reserve University, Cleveland, OH, USA).
Application of FT-IR and NMR to Epoxy Resins (Part II, Vol. 75).

R. J. Morgan (Lawrence Livermore National Laboratory, Livermore, CA, USA).
Structure-Properties Relations of Epoxies Used as Composite Matrices (Part I, Vol. 72).

E. F. Oleinik (Institute of Chemical Physics, Academy of Sciences of USSR, Moscow, USSR).
Structure and Properties of Epoxy-Aromatic Amine Networks in the Glassy State (Part IV, Vol. 80).

B. A. Rozenberg (Institute of Chemical Physics, Academy of Sciences of USSR, Moscow, USSR).

Kinetics, Thermodynamics and Mechanism of Reactions of Epoxy Oligomers with Amines (Part II, Vol. 75).
S. D. Senturia and N. F. Sheppard (Massachusetts Institute of Technology, Cambridge, MA, USA).
Dielectric Analysis of Epoxy Cure (Part IV, Vol. 80).
R. G. Schmidt and J. P. Bell (University of Connecticut, Storrs, CT, USA).
Epoxy Adhesion to Metals (Part II, Vol. 75).
E. M. Yorkgitis, N. S. Eiss, Jr., C. Tran, G. L. Wilkes and J. E. Mc Grath (Virginia Polytechnic Institute, Blacksburg, VA, USA).
Siloxane Modified Epoxy Resins (Part I, Vol. 72).

The editor wishes to express his gratitude to all contributors for their cooperation.

Prague, January 1986
Karel Dušek
Editor

# Table of Contents

**Dielectric Analysis of Thermoset Cure**
S. D. Senturia, N. F. Sheppard, Jr. . . . . . . . . . . . 1

**Epoxy-Aromatic Amine Networks in the Glassy State. Structure and Properties**
E. F. Oleinik . . . . . . . . . . . . . . . . . . . . . 49

**Void Growth and Resin Transport During Processing of Thermosetting — Matrix Composites**
J. L. Kardos, M. P. Duduković, R. Dave . . . . . . . . 101

**Physical Aging in Epoxy Matrices and Composites**
E. S.-W. Kong . . . . . . . . . . . . . . . . . . . . 125

**Curing Mechanisms and Mechanical Properties of Cured Epoxy Resins**
T. Kamon, H. Furukawa . . . . . . . . . . . . . . . 173

**Author Index Volumes 1–80** . . . . . . . . . . . . . 203

**Subject Index** . . . . . . . . . . . . . . . . . . . . 215

# Dielectric Analysis of Thermoset Cure

Stephen D. Senturia
Massachusetts Institute of Technology, Cambridge, MA 02139, USA

Norman F. Sheppard, Jr.
Massachusetts Institute of Technology, Cambridge MA 02139, USA

*All dielectric measurements involve the determination of the electrical polarization and conduction properties of a sample subjected to a time-varying electric field. Section 2 addresses dielectric measurement methods, the various instruments and electrodes, and their calibrations. Section 3 examines the microscopic mechanisms giving rise in the observed microscopic dielectric properties, and Section 4 explores in detail the effects of temperature and cure on these properties. Finally, Section 5 contains a selected bibliography of applications of dielectric analysis to the study of thermoset cure.*

| | |
|---|---:|
| **1 Introduction** | 3 |
|   1.1 Historical Perspective | 3 |
| **2 Dielectric Measurement Methods** | 4 |
|   2.1 Admittance Measurements | 4 |
|   2.2 Electrode Geometries and Their Calibrations | 8 |
|     2.2.1 Definitions | 8 |
|     2.2.2 Parallel Plate Electrodes | 8 |
|     2.2.3 Comb Electrodes | 9 |
|     2.2.4 Other Electrodes | 12 |
|   2.3 Measurement Equipment | 12 |
|     2.3.1 Capacitance and Impedance Bridges | 12 |
|     2.3.2 Microdielectrometry | 14 |
| **3 Microscopic Mechanisms** | 14 |
|   3.1 Bulk Effects | 14 |
|     3.1.1 Ionic Conductivity | 15 |
|     3.1.2 Dipole Orientation | 16 |
|   3.2 Interface Effects | 20 |
|     3.2.1 Electrode Polarization | 20 |
|     3.2.2 Blocking and/or Release Layers | 23 |
|     3.2.3 Fibers and Fillers | 24 |
| **4 Effects of Temperature and Cure** | 25 |
|   4.1 General Issues | 25 |
|   4.2 Relation with Chemical Kinetics | 28 |

4.3 Relaxed Permittivity . . . . . . . . . . . . . . . . . . . . 29
   4.4 Dipolar Relaxations . . . . . . . . . . . . . . . . . . . . 32
   4.5 Conductivity . . . . . . . . . . . . . . . . . . . . . . . 36

**5 Applications** . . . . . . . . . . . . . . . . . . . . . . . . . 40

**6 Nomenclature** . . . . . . . . . . . . . . . . . . . . . . . . 42

**7 References** . . . . . . . . . . . . . . . . . . . . . . . . . 43

# 1 Introduction

## 1.1 Historical Perspective

Measurements of dielectric properties have been used to monitor chemical reactions in organic materials for more than fifty years. In 1934, Kienle and Race [1] reported the use of dielectric measurements to study polyesterification reactions. Remarkably, many of the major issues that are the subject of this review were identified in that early paper: the fact that ionic conductivity often dominates the observed dielectric properties; the equivalence between the conductivity measured with both DC and AC methods; the correlation between viscosity and conductivity early in cure; the fact that conductivity does not show an abrupt change at gelation; the possible contribution of orientable dipoles and sample heterogeneities to measured dielectric properties; and the importance of electrode polarization at low frequencies.

Between 1934 and 1958, the literature is sparse, but significant. Manegold and Petzoldt, writing in Germany in 1941 [2], cited a 1939 translation [3] of the abstract of a 1937 Russian article (which we were unable to obtain) by Lomakin and Gussewa [4] in which a correspondence was reported between the viscosity and both the electrical conductivity and index of refraction during the cure of phenol-formaldehyde resins. The Manegold and Petzoldt paper focuses on the conductivity changes during phenolic cure, examining the effects of both stoichiometry and catalyst variation. Their experiments included simultaneous temperature and conductivity measurements, permitting them to separate the intrinsic temperature dependence of the conductivity from the reaction-induced changes. They further demonstrated with the catalyst variation that the rate of change of conductivity after the onset of reaction varied with the reaction rate. Fineman and Puddington in 1947 [5, 6] extended these conductivity studies to resorcinol-formaldehyde resins and to a commercial polyester, adding correlations with density measurements. Based on an observed similarity between the data for the two resins, only one of which loses water during cure, they concluded that the conductivity changes observed during cure were due to the changing molecular network, and not simply to water loss.

Since 1958, an extensive experimental literature has developed, primarily on epoxies, and, to a lesser extent, on polyesters, polyimides, phenolics, and other resins. Work in this area has been greatly stimulated by the increasing importance of thermosets as matrix resins in fiber reinforced composites, and by significant improvements in instrumentation and measurement methods.

One problem that has plagued the field has been the overwhelmingly empirical nature of the research, hampered by inadequate models with which to interpret the data. Furthermore, in only a few cases have the dielectric measurements been quantitatively coupled with measurements of other properties of interest, and often, experimental details that turn out to be important in hindsight, were inadvertently overlooked during the original work. As a result, many well intentioned experiments have somehow failed to provide that cumulative insight into fundamental issues that must ultimately accompany the successful scientific application of a measurement method. Thus, a major goal of this article has been to present, within a single source, a unified review of basic dielectric properties, the methods used to measure those properties, key experimental artifacts that must be understood by the investigator, the physical origins of those artifacts, and a survey of the current published literature with emphasis on the correlations that can be found between the dielectric properties and other physical properties of the curing system.

## 2 Dielectric Measurement Methods

### 2.1 Admittance Measurements

Dielectric measurements are performed by placing a sample of the material to be studied between two conducting electrodes, applying a time-varying voltage between the electrodes, and measuring the resulting time-varying current (an "admittance" measurement). The applied voltage establishes an electric field in the sample. In response, the sample can become electrically polarized (the conventional "dielectric" response) and can also conduct net charge from one electrode to the other. Both dielectric polarization and conduction give rise to currents which can vary enormously during a cure.

It is useful to separate the phenomena associated with the *measurement* from the interpretation of those phenomena in terms of the dielectric properties of the sample. Figure 1 shows a "black-box" view of the sample, in which the details of the shape of the electrodes and the precise properties of the material between the electrodes are not known; only the external terminals of the electrodes are available. It is assumed that the "apparatus" of Fig. 1 applies a time-varying voltage v(t) between the terminals of the sample, and has the capability of determining either the time-varying current i(t), or the time-varying net charge Q(t), which is the time integral of the current. The discussion that follows deals with the current; an equivalent set of ideas based on charge is implied.

The interpretation of dielectric measurements assumes that the sample behavior can be represented by a *linear, time-invariant admittance*. The meaning of each of these terms is examined in turn.

*Linearity*, when dealing with time-varying signals in circuits that have energy-storage elements (such as capacitors, which are made from electrodes with a dielectric medium between them), is not the same as simple proportionality. A sufficient condition for linearity can be expressed in terms of the superposition property, as follows: a sample is linear if, given that $i_1(t)$ is the response to a particular waveform $v_1(t)$, and $i_2(t)$ is the response to a second waveform $v_2(t)$, then the response to the waveform $av_1(t) + bv_2(t)$ (where a and b are constants) is $ai_1(t) + bi_2(t)$. (A small detail: all linear samples obey the superposition condition above provided that the net polarization charge before applying the voltage waveforms is zero. If a sample

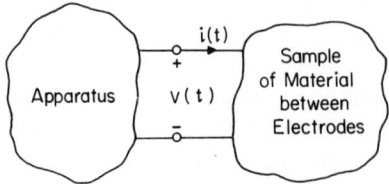

**Fig. 1.** "Black-box" view of dielectric measurement. Apparatus applies a time-varying voltage v(t) to the electrodes and measures the time-varying current i(t)

has residual polarization from previous experiments, even linear samples may fail to obey superposition.)

Whether or not a given sample actually is linear, depends on the magnitude of the applied voltage. All dielectrics experience catastrophic breakdown at electric fields of the order of $10^6$ volts/cm. Most dielectric measurements are made with applied voltages of the order of 1 volt, with electrode spacings ranging from tens of μm to several mm, resulting in electric fields well below breakdown. The dielectric portion of the response, therefore, is usually linear. The conduction portion of the response is often due to ions. At high enough applied voltages, electrochemical interactions at the electrodes can occur which can lead to nonlinear conduction characteristics. There can also be intrinsically nonlinear conduction mechanisms within the medium at electric fields in excess of about $10^5$ V/cm. Coln has recently looked carefully for nonlinearities in liquid epoxy resins prior to cure (where ion conduction is most important) and has found no significant nonlinear effects under normal measurement conditions [7]. It is reasonable to assume, therefore, that any sample used in conventional dielectric measurements can be considered linear.

*Time-invariance* presents a problem. Since the purpose of the measurement is to follow *changes* in dielectric and conducting properties of the sample, strictly speaking, the sample is usually not time-invariant. Yet, time-invariant circuit models are always used to interpret the results of the measurement. The justification for this practice is to assume that the sample properties change insignificantly during the interval required to make a single measurement. The degree to which this may or may not be true during actual cure studies has received relatively little attention to date, because only recently have there been commercial instruments available for cure studies below 10 Hz, where it might be possible to demonstrate explicit reaction-rate dependent effects.

Dielectric measurements on curing systems are usually done with sinusoidal excitations at specific frequencies of interest. The frequency-dependent dielectric response of a linear *time-invariant* sample can be obtained by applying a step-change in voltage, measuring the resulting current waveform at a series of time intervals following the step, then computing the frequency-dependent dielectric properties from the Fourier transform of the current waveform [8]. The potential advantage of this approach is the use of digital signal processing methods to reduce the time needed to obtain data at ultra-low frequencies (to $10^{-4}$ Hz), but in practice, the duration of the step waveform must be short enough to keep the sample time-invariant throughout one measurement cycle. The total time for a measurement, in this case, is determined, first, by the length of time one must hold the voltage at some initial voltage (e.g., zero) before applying the step (to assure zero initial polarization charge), and second, by the length of time one must hold the voltage at its non-zero value to obtain enough data for the lowest frequency of interest. For typical thermoset cures, reaction rates usually limit the lowest useful frequency to about 0.1 Hz.

The *admittance* of a linear time-invariant sample can now be defined. In the sinusoidal steady state at angular frequency ω, both v(t) and i(t) are sinusoids having some phase difference φ (see Fig. 2).

$$v(t) = V_0 \cos(\omega t) \qquad (2\text{-}1)$$

$$i(t) = I_0 \cos(\omega t + \varphi) \qquad (2\text{-}2)$$

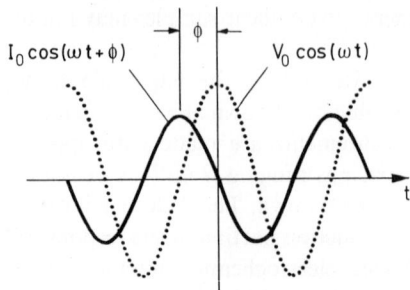

**Fig. 2.** Sinusoidal voltage and current waveforms having a phase difference ∅

where $V_0$ and $I_0$ are the real amplitudes of $v(t)$ and $i(t)$, respectively. These waveforms can be expressed in complex exponential notation, as follows:

$$v(t) = \text{Re}\{Ve^{j\omega t}\} \tag{2-3}$$

$$i = \text{Re}\{Ie^{j\omega t}\} \tag{2-4}$$

where $j$ is $(-1)^{1/2}$, and where V and I are the *complex amplitudes* of $v(t)$ and $i(t)$, respectively. When expressed in this form, V and I have the values

$$V = V_0 \tag{2-5}$$

$$I = I_0 e^{j\varphi} \tag{2-6}$$

The ratio of I to V is defined as the *admittance* $Y(\omega)$ of the sample, and is written as frequency-dependent to emphasize its implicit dependence on the frequency-dependent properties of the medium.

$$Y(\omega) = I/V \tag{2-7}$$

Substituting and expanding yields

$$Y(\omega) = j(I_0/V_0)\sin\varphi + (I_0/V_0)\cos\varphi \tag{2-8}$$

In this Equation, the magnitude of the admittance $(I_0/V_0)$ and the phase $\varphi$ are both frequency-dependent quantities. Nevertheless, at some particular frequency, it is possible to construct an equivalent circuit consisting of a capacitor $C_x(\omega)$ in parallel with a resistor $R_x(\omega)$, as shown in Fig. 3. (For clarity in this introductory discussion, the subscript "x" is used to denote "experimentally measured quantities" to distinguish them from bulk properties of materials, which are given without subscripts.) Until the electrode geometry is specified, the capacitor and resistor have no direct physical interpretation in terms of dielectric properties of the sample; they simply provide a convenient way of describing the results of the measurement. Electrode geometries are considered in the following section. For the parallel

$R_x - C_x$ circuit of Fig. 3, the admittance is written

$$Y(\omega) = j\omega C_x(\omega) + 1/R_x(\omega) \tag{2-9}$$

which, by comparison with Eq. (2-8), yields

$$C_x(\omega) = \frac{I_0 \sin \varphi}{\omega V_0} \tag{2-10}$$

and

$$R_x(\omega) = \frac{V_0}{I_0 \cos \varphi} \tag{2-11}$$

Another quantity also used in admittance measurements is the *dissipation factor* D, which is useful because it depends only on the phase of the admittance, not on its magnitude. The dissipation factor of a parallel $R_x - C_x$ circuit is defined as

$$D = \frac{1}{\omega R_x C_x} \tag{2-12}$$

which from Eqs. (2-10) and (2-11), yields

$$D = \cotan \varphi \tag{2-13}$$

The final quantity to be defined has been the source of much confusion. The *loss tangent* of the *sample* is the same as the dissipation factor defined above; however, the loss tangent of the *medium* is a dielectric property. To distinguish between them, we shall refer to $\tan \delta_x$ as the sample loss tangent, having the value

$$\tan \delta_x = D \tag{2-14}$$

Therefore, $\delta_x$ is $\pi/2 - \varphi$. The loss tangent of the medium, which we shall refer to as $\tan \delta$ without a subscript, is discussed in Section 3.

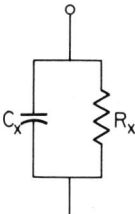

**Fig. 3.** Equivalent circuit for the admittance of the electrodes and sample of Fig. 1.

## 2.2 Electrode Geometries and Their Calibrations

### 2.2.1 Definitions

The *dielectric permittivity* of a medium measures the polarization of the medium per unit applied electric field. The permittivity of free space, $\varepsilon_0$ has the value $8.85 \times 10^{-14}$ Farads/cm. Throughout this paper, we shall express the dielectric properties of a medium relative to $\varepsilon_0$. That is, the permittivity of a medium is written $\varepsilon'\varepsilon_0$, where $\varepsilon'$ is the *relative permittivity*, also referred to as the *dielectric constant* of the medium (but remember that the dielectric "constant" changes with temperature and cure).

The *dielectric loss factor* arises from two sources: energy loss associated with the time-dependent polarization, and bulk conduction. The loss factor is written $\varepsilon''\varepsilon_0$, where $\varepsilon''$ is the relative loss factor.

The *loss tangent* of a *medium*, denoted by tan $\delta$, is defined as the ratio $\varepsilon''/\varepsilon'$ (see also Section 2.1 for the loss tangent of a sample, tan $\delta_x$).

### 2.2.2 Parallel Plate Electrodes

Parallel plates are often used as electrodes in cure studies; their use in this application is first cited by Manegold and Petzold [2] and later by Aukward and Warfield [9]. Figure 4 illustrates parallel conductors of area A separated by spacing L. The sample medium is placed between the plates, or the plates can be fully immersed in the medium. Normally, one designs the plate spacing to be much less than the plate size, which minimizes the effects of fringing fields and leads to a very simple calibration. After accounting for any cabling admittances (see Sect. 2.3.1), the components of the equivalent circuit of Fig. 3 have the values:

$$C_x = \frac{\varepsilon'_x \varepsilon_0 A}{L} \tag{2-15}$$

and

$$R_x = \frac{L}{\omega A \varepsilon''_x \varepsilon_0} \tag{2-16}$$

where $\varepsilon'_x$ and $\varepsilon''_x$ denote the experimental values that are inferred for $\varepsilon'$ and $\varepsilon''$, respectively, and where, to be consistent with the units selected for $\varepsilon_0$, linear dimensions are in cm.

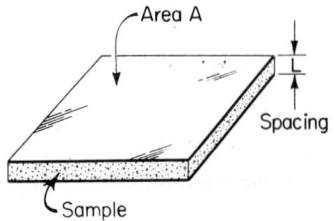

**Fig. 4.** Parallel plate electrodes

For a homogeneous dielectric medium, in the absence of interface effects (see Sect. 3.2), the experimental quantities $\varepsilon'_x$ and $\varepsilon''_x$ are equal to the bulk properties $\varepsilon'$ and $\varepsilon''$, and the experimental loss tangent becomes

$$\tan \delta_x = \tan \delta \tag{2-17}$$

That is, for a parallel plate structure containing a homogeneous medium and having no interface effects, the phase angle of the sample admittance determines the loss tangent of the medium, regardless of the plate spacing and area (within wide limits). Of course, if calibrated measurements of either $\varepsilon'$ or $\varepsilon''$ are needed (and, as will be shown, they usually are), then it is necessary to measure and control both plate spacing and area to obtain quantitative results. The "cell constant" calibration method, in which the ratio A/L in Eq. (2-15) is implicitly determined by measuring the electrode capacitance in air ($\varepsilon' = 1.0$), is not sufficient for calibration in the presence of interface effects; L must be known separately (see Sect. 3.2).

The primary advantages of parallel plate electrodes lie in the ease with which measured data can be interpreted, and in the fact that they can be made from almost any conductor. The primary disadvantages are the need the control plate spacing and area if quantitative measures of either $\varepsilon'$ or $\varepsilon''$ are to be obtained, and the instrumental difficulties associated with making admittance measurements at frequencies below about 100 Hz (see Sect. 2.3). Furthermore, in cure experiments, matrix resins typically go through significant dimensional changes either due to temperature, to reaction-induced contraction, or to applied pressure, and it has proved difficult to maintain the calibration of parallel plates in these cases. This has led to almost complete reliance on the measured $\tan \delta_x$ in following cure, but as is shown in Section 3.2, interface effects lead to complex artifacts in $\tan \delta_x$. As a result of these disadvantages, the use of parallel plates for quantitative studies is decreasing relative to the use of comb electrodes.

### 2.2.3 Comb Electrodes

The use of interdigitated, or *comb* electrodes for cure measurement was first reported by Armstrong [10]. The metal electrodes, illustrated schematically in Fig. 5a, are typically fabricated on an insulating substrate using photopatterning. The substrate can be a ceramic, a thin plastic film, a sheet of epoxy-glass composite, or even a silicon integrated circuit. The Cross-Section in Fig. 5b illustrates the case where the insulating substrate is much thicker than the spacing between fingers. The calibration of the device is found by determining the admittance between the arrays of parallel stripe electrodes located at the interface between one medium having the

 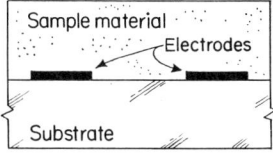

**Fig. 5a and b. a)** Top view of comb electrodes. **b)** Cross-Section where the insulating substrate is much thicker than the electrode spacing

a  b

ε' and ε" values of the sample, the other medium having the ε' and ε" values of the substrate. In general, this calibration cannot be calculated for exact device geometries without numerical methods. If the interdigitated portion of the electrodes is sufficiently large, the structure can be approximated by an infinite periodic comb structure, for which both analytical and simplified numerical calibrations can be obtained [11,12]. The cell constant calibration method fails for comb electrodes even in the absence of interface effects because of the relatively large component of fringing field in the substrate.

The advantages of comb electrodes lie in the *reproducibility* of the calibration (which permits separation of ε' and ε"), and in the ease with which they can be placed into a variety of structures, such as adhesive joints and fiber-reinforced laminates. The calibration depends on the electrode size and spacing and on the dielectric properties of the substrate, all of which can be made quite reproducible with photopatterning technology. Furthermore, the calibration is relatively insensitive to temperature or pressure changes, a major advantage compared with parallel plates. Finally, when combined with suitable blocking or release layers (see Sect. 3.2), comb electrodes can be used in fiber-reinforced laminates containing graphite fibers, which are such good conductors that they interfere substantially with parallel plate measurements. The major disadvantage of comb electrodes is that they are much less sensitive than parallel plates of comparable dimensions. As a result, their primary application is in those materials where the loss factor is dominated by conductivity effects and where the conductivity is relatively large, i.e., greater than about $10^{-8}$ $(\Omega\text{-cm})^{-1}$ (see Sect. 3.1). This class of materials is sufficiently large to provide a large domain of application for comb electrodes.

An important variation of the comb electrode approach is found in microdielectrometry [13,14]. The microdielectrometer sensor combines a comb electrode with a pair of field-effect transistors in a silicon integrated microcircuit to achieve sensitivities comparable with parallel-plate electrodes, but retaining the reproducible calibration features of the comb. The Cross-Section of Fig. 6 shows the electrodes separated from a conducting ground plane (the silicon substrate) by a silicon dioxide insulator whose thickness is much less than the electrode spacing. One of the electrodes is driven with a signal, the other is connected to the input gate of one of the field-effect transistors, and except for a capacitance $C_L$ between this electrode and the ground plane, is electrically floating. The capacitance $C_L$ integrates the current reaching the floating electrode through the comb electrode admittance $Y(\omega)$, and develops a voltage which depends on the charge rather than the current. Therefore, instead of providing a direct measurement of $Y(\omega)$, the microdielectro-

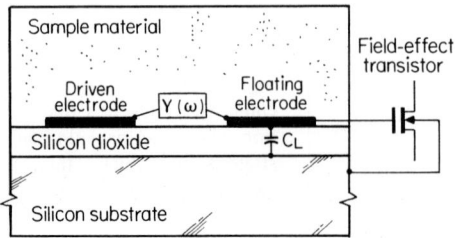

**Fig. 6.** Cross-Section of microdielectrometer sensor. The silicon dioxide insulator is much thinner than the electrode spacing. $Y(\omega)$ is the comb electrode admittance; $C_L$ is the capacitance between the floating electrode and the substrate.

meter measures this voltage, or, equivalently, the complex transfer function $H(\omega)$ defined as follows:

$$H(\omega) = \frac{Y(\omega)}{1 + j\omega C_L Y(\omega)} \qquad (2\text{-}18)$$

Like the admittance $Y(\omega)$, the transfer function $H(\omega)$ has a magnitude and phase which can be interpreted in terms of an assumed homogeneous dielectric medium with permittivity $\varepsilon'$ and loss factor $\varepsilon''$ [12]. Figure 7 illustrates such a calibration as a contour plot. Given the magnitude and phase of $H(\omega)$, the corresponding $\varepsilon'$ and $\varepsilon''$ values can be determined. Furthermore, as was the true for the parallel plates and simple comb electrodes, calibrations like that of Fig. 7 implicitly assume a homogeneous dielectric medium and no interface effects. In practice, however, interface effects are almost always observed at the early stages of thermoset cure (see Sect. 3.2).

A schematic view of a microdielectrometer sensor is shown in Fig. 8 and illustrates the electrode array, the field-effect transistors and a silicon diode temperature indicator [15] which functions as a moderate accuracy ($\sim 2$ °C) thermometer between room temperature and 250 °C. The sensor is used either by placing a small sample of resin over the electrodes, or by embedding the sensor in a reaction vessel or laminate. Since all dielectric and conductivity properties are temperature dependent, the ability to make a temperature measurement at the same point as the dielectric measurement is a useful feature of this technique.

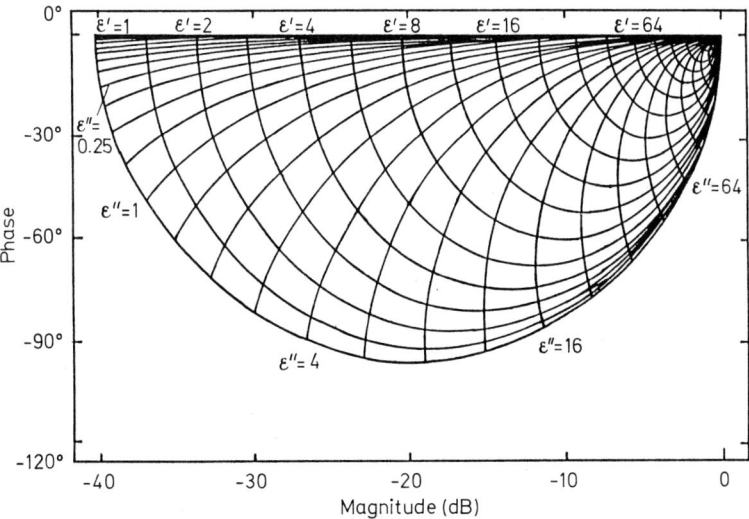

**Fig. 7.** Calibration of the microdielectrometer sensor, showing contours of constant permittivity and loss factor as a function of the magnitude and phase of the transfer function $H(\omega)$. (Reprinted from Ref. [7] with permission of IEEE, © 1985 IEEE.)

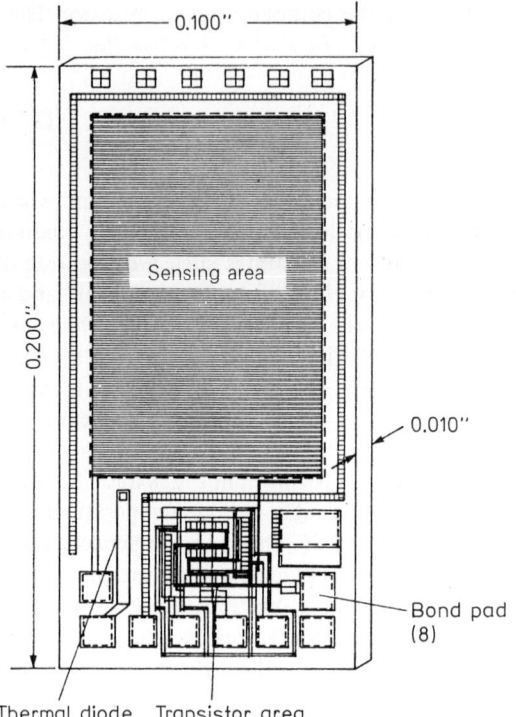

**Fig. 8.** Schematic top view of actual microdielectrometer sensor, illustrating the comb electrode structure, field effect transistors and thermal diode temperature indicator. (Reprinted with permission of Micromet Instruments, Inc.)

2.2.4 Other Electrodes

Many electrode patterns have been used for cure studies. Kienle and Race [1] used parallel cylindrical conductors immersed in the medium. Coaxial cylinder electrodes, with the sample placed between the two cylinders, were first used by Fineman and Puddington [5,6], and later by Aukward, Warfield, and Petree [16]. The coaxial electrodes, like the parallel plates, have a relatively simple calibration, but are tedious to construct reproducibly for each thermoset cure experiment. Generally, any electrode configuration can be used for observing trends, but electrode size and location must be reproducibly controlled if calibrated measurements are desired.

## 2.3 Measurement Equipment

2.3.1 Capacitance and Impedance Bridges

Whether using parallel plate or comb electrodes, the basic measurement involves determining the admittance betwen the electrodes under sinusoidal steady-state conditions (the lone exception is the microdielectrometer, discussed in Sect. 2.3.2). Many early studies were made with general-purpose capacitance bridges designed for measurements of capacitors with small loss tangents. Because thermoset resins become relatively conductive when they are heated, particularly early in cure, such capacitance bridges could not be used without the addition of an insulating (or blocking) layer between at least one of the electrodes and the sample to lower the overall loss tangent of the sample to within the instrument range. However, this significantly modifies the calibration (see Sect. 3.2).

The first commercial admittance bridge specifically directed toward resin-cure applications was introduced by Tetrahedron Associates in 1969 [17]. Within the past decade, a variety of general-purpose instruments suitable for resin-cure applications have been introduced. Examples from major U.S. manufacturers include the Hewlett-Packard HP4192A Low Frequency Impedance Analyzer [18], and the GenRad 1689 Digibridge [19]. Either can provide measurements of both $R_x$ and $C_x$ over a range of frequencies (11 Hz–100 kHz for the GenRad, and 5 Hz–13 MHz for the HP). Both are equipped with the equivalent of the standard IEEE 488 computer bus, which facilitates interfacing and control from a variety of data-logging computers or desktop calculators.

The effectiveness of these instruments for dielectric cure studies depends on sensitivity and accuracy. The sensitivity is related to the minimum resolvable phase angle, which for general cure studies, should ideally be less than about 0.10. Unfortunately, actual sensitivity in use depends strongly on the measurement frequency, on the admittance of the sample, on the details of the cabling and shielding, and on the electrical noise level of the environment. Therefore, analysis of published sensitivity specifications is difficult. It is easier to evaluate intrinsic instrument accuracy, which can be expressed in terms of either the $\tan \delta_x$ accuracy or the conductivity accuracy. An example is useful.

A pair of 1 cm² area plates spaced apart by 0.25 mm and filled with a resin having a permittivity of 10 (a typical value early in cure) has a capacitance of about 35 pF. The HP4192A has a $\tan \delta_x$ accuracy of 0.002 when measuring 35 pF at 1000 Hz [18], which is satisfactory for most resin studies at that frequency. However, the $\tan \delta_x$ accuracy of the HP4192A degrades to about 0.05 at 5 Hz, which limits the smallest conductivity that can be measured. In the final stages of typical cures, $\varepsilon'$ approaches a value of 4–5, while $\varepsilon''$ approaches a value that depends on frequency. At low frequencies, the $\varepsilon''$ value is usually dominated by ionic conductivity, denoted by $\sigma$ (see Sect. 3.1.1). In this case, the resistance $R_x$ is $L/\sigma A$, which when combined with Eq. (2-16) yields

$$\sigma = \omega \varepsilon_0 \varepsilon' \tan \delta_x \qquad (2\text{-}19)$$

If at 5 Hz, $\tan \delta_x$ is only accurate to 0.05, and using $\varepsilon' = 4$, the corresponding accuracy for $\sigma$ is $1.9 \times 10^{-13}$ $(\Omega\text{-cm})^{-1}$. For many materials including epoxies, it is desirable to measure $\sigma$ values into the $10^{-14}$ $(\Omega\text{-cm})^{-1}$ range, or lower.

Additional problems are introduced by the overall admittance level and by cabling and shielding issues. In Fig. 9, the parallel-plate sample is connected to a meter

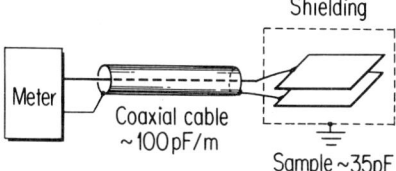

Fig. 9. Schematic diagram of admittance measurement

by a coaxial cable (~100 pF/meter). A 2-meter cable introduces a capacitance of 200 pF in parallel with the sample, which must be subtracted from the measured capacitance to determine the sample capacitance. Clearly, when the cable capacitance is comparable to the sample capacitance, sensitivity and accuracy are reduced. Furthermore, unless the cable and electrodes are shielded, pickup at 60 Hz and its harmonics can degrade measurements at or near those frequencies. Finally, at frequencies below 100 Hz, the magnitude of the sample admittance becomes small ($\sim 2 \times 10^{-8}$ $\Omega^{-1}$ at 100 Hz, and decreasing proportional to $\omega$ at lower frequencies), and becomes comparable to stray or leakage admittances within the instrument. In practice, it has proved difficult to operate admittance-measuring instruments below about 10 Hz, and even an instrument with an ideal tan $\delta_x$ sensitivity as low as 0.001, if restricted by cabling problems to a useful frequency range of 10 Hz or above, cannot measure a conductivity smaller than about $1.5 \times 10^{-14}$ $(\Omega\text{-cm})^{-1}$.

### 2.3.2. Microdielectrometry

Microdielectrometry was introduced as a research method in 1981 [14] and became commercially available in 1983 [20]. The microdielectrometry instrumentation combines the pair of field-effect transistors on the sensor chip (see Sect. 2.2.3) with external electronics to measure the transfer function $H(\omega)$ of Eq. (2-18). Because the transistors on the sensor chip function as the input amplifier to the meter, cable admittance and shielding problems are greatly reduced. In addition, the use of a charge measurement rather than the admittance measurement allows the measurements to be made at arbitrarily low frequencies. As a matter of practice, reaction rates in cure studies limits the lowest useful frequency to about 0.1 Hz; however, pre-cure or post-cure studies can be made to as low as 0.005 Hz. Finally, the differential connection used for the two transistors provides first-order cancellation of the effects of temperature and pressure on the transistor operation. The devices can be used for cure measurements to 300 °C, and at pressures to 200 psi.

As described in Section 2.2.3, the microdielectrometer calibration [12] is similar to that of comb electrodes. Based on the accuracy of the amplitude and phase measurement electronics, the $\varepsilon''$ sensitivity of the microdielectrometer is about 0.01 [7], which for a medium having a dielectric permittivity of 4 corresponds to a tan $\delta$ sensitivity of less than 0.003. At a frequency of 0.1 Hz, an $\varepsilon''$ sensitivity of 0.01 corresponds to a conductivity sensitivity of about $1 \times 10^{-16}$ $(\Omega\text{-cm})^{-1}$. However, the accuracy of the microdielectrometry calibration at these conductivity levels has not been rigorously established.

# 3 Microscopic Mechanisms

## 3.1 Bulk Effects

This Section examines the dielectric and conduction mechanisms in bulk materials, assuming that the medium is linear (at the applied electric field strength) and homogeneous. Effects of interfaces and inhomogeneities are discussed in Section 3.2. Additional discussion can be found in basic texts [21-23].

Two bulk effects are considered in the following Sections: ionic conductivity, and molecular dipole orientation. It is also necessary to introduce the so-called "infinite-frequency" dielectric polarization which provides the baseline against which to measure the other effects. The permittivity $\varepsilon'$ is written schematically as

$$\varepsilon' = \varepsilon_\infty + \varepsilon'_d \qquad (3\text{-}1)$$

where $\varepsilon_\infty$ is the baseline permittivity, and $\varepsilon'_d$ is the additional permittivity attributed to dipole orientation. For the purpose of this review, $\varepsilon'_d$ is associated with the major dipolar relaxation at the glass transition (the $\alpha$-relaxation); it depends on frequency, temperature, and cure. The frequency-dependence is examined in Section. 3.1.2; the temperature and cure dependences are discussed in Section 4.

The definition of $\varepsilon_\infty$ depends somewhat on the temperature and/or frequency range used for the experiments. Typical dielectric studies take place at frequencies below about 1 MHz. At temperatures below the glass transition, there can still be dipolar contributions to $\varepsilon'$ at these frequencies from limited motions of polar groups. However, at sufficiently low temperatures (and/or high frequencies), these groups lose the ability to align with an applied electric field, resulting in additional decreases in $\varepsilon'$ (the $\beta$- and lower-temperature relaxations). This review does not address either the low-temperature properties of resins, or the details of these relaxations. Hence, such dipolar effects are lumped into $\varepsilon_\infty$. Within the temperature range at which typical cures take place, $\varepsilon_\infty$ is found not to depend significantly on frequency, temperature, or cure.

### 3.1.1 Ionic Conductivity

The importance of ionic conductivity in curing resins has been recognized since the earliest work [1,2]. In epoxies, Fava [24] proposed that sodium and chloride ions are the particular species involved, the origin of the ions being the reaction used to produce the starting materials. (The reaction of epichlorhydrin with Bisphenol A to make the diglycidyl ether of Bisphenol A (DGEBA) produces HCl as a byproduct which is subsequently neutralized with alkali [25].) Even after treatment to remove NaCl, there is residual chloride ion present in commercial DGEBA resins at concentrations typically on the order of tens of ppm [26], and corresponding concentrations of cations. These impurities actually provide a remarkably useful probe of the resin system.

If the electric field within the resin is E, the $i^{th}$ species of ion will acquire an average drift velocity $v_i$. The assumed linearity of the medium implies that $v_i$ is proportional to E. The proportionality constant is called the *mobility* of the ion, for which we use the symbol $u_i$.

$$v_i = u_i E \qquad (3\text{-}2)$$

If there are $N_i$ ions of species i per unit volume, with a charge magnitude of $q_i$ on the ith ion, the ionic conductivity $\sigma$ can be expressed as

$$\sigma = \sum_i q_i N_i u_i \qquad (3\text{-}3)$$

The relation between the mobility of the ion and the properties of the resin can be *qualitatively* examined with the aid of Stoke's law for the drift of a spherical object in a viscous medium (see, for example, Ref. [27]). The mobility of a sphere of radius $r_i$ embedded in a medium of viscosity $\eta$ and subjected to a force $q_i E$ is

$$u_i = \frac{q_i}{6\pi\eta r_i} \qquad (3\text{-}4)$$

In this simple model, the mobility, and hence $\sigma$, varies as $1/\eta$; equivalently, the quantity $1/\sigma$, called the *resistivity* and denoted $\varrho$, is nominally proportional to viscosity. It must be emphasized, however, that this Stoke's law approach is an oversimplification which fails completely as a curing resin approaches gelation. As discussed in detail in Section 4, the ion mobility in a resin depends primarily on the mobilities of the polymer segments. At gelation, the bulk viscosity becomes infinite because of the formation of a macroscopic molecular network. However, the resistivity remains finite because polymer segments comparable in size to the ions are still mobile. Well before gelation, the resistivity and viscosity *are* tightly correlated because both the viscosity and ion mobility have similar dependences on polymer segment mobility.

Ionic conductivity has another important implication. The resin system acts like an electrolyte; thus, all of the electrode polarization effects that can be observed in conventional electrolytes can also be observed in resins. The effect of electrode polarization is discussed in Section 3.2.

Conductivity effects give rise to a $1/\omega$ frequency dependence in $\varepsilon''$, as shown below

$$\varepsilon'' = \frac{\sigma}{\omega\varepsilon_0} + \varepsilon_d'' \qquad (3\text{-}5)$$

where $\varepsilon_d''$ is the contribution to $\varepsilon''$ from losses arising from dipolar orientation. This frequency dependence is examined further in the following Section.

3.1.2. Dipole Orientation

Figure 10 illustrates in highly schematic form the alignment of molecular dipoles in an applied electric field. The dipoles in a curing resin are embedded in a viscous medium, and are hindered by attachment to a growing network. The orientation process will require a characteristic time, called the *dipole relaxation time* and denoted by $\tau_d$. During a typical cure reaction, $\tau_d$ is short early in cure, and becomes large when the resin vitrifies. Because of the hindering mechanism, there is energy loss associated with the orientation process.

The following notation is used to describe the dipolar quantities:

$\varepsilon_u$: The "unrelaxed" permittivity, equivalent to $\varepsilon_\infty$;
$\varepsilon_r$: The "relaxed" permittivity, equal to the bulk permittivity when molecular dipoles align with the electric field to the maximum extent possible at the sample temperature.

The simplest model of hindered dipole orientation is due to Debye [28], and assumes a single relaxation time for all molecular species. The Debye model, plus a term to account for ionic conductivity, leads to the following illustrative expressions for $\varepsilon'$ and $\varepsilon''$:

$$\varepsilon' = \varepsilon_u + \frac{\varepsilon_r - \varepsilon_u}{1 + (\omega \tau_d)^2} \qquad (3\text{-}6)$$

and

$$\varepsilon'' = \frac{\sigma}{\omega \varepsilon_0} + \frac{(\varepsilon_r - \varepsilon_u)\omega \tau_d}{1 + (\omega \tau_d)^2} \qquad (3\text{-}7)$$

Figure 11a illustrates the frequency dependence of $\varepsilon'$ for Eq. (3-6). Note that $\varepsilon'$ is midway between $\varepsilon_u$ and $\varepsilon_r$ when $\omega = 1/\tau_d$. The corresponding plots for $\varepsilon''$ are more complex, because one must assess the relative contributions of $\sigma$ and the dipole loss. The simplest case is for $\sigma = 0$ (Fig. 11b), where the characteristic dipolar loss peak of amplitude $(\varepsilon_r - \varepsilon_u)/2$ is observed at frequency $\omega = 1/\tau_d$. For non-zero $\sigma$, however, the $1/\omega$ dependence of $\varepsilon''$ greatly distorts the $\varepsilon''$ curve from the ideal Debye peak. Log-log scales are helpful, as illustrated in Fig. 12. The $\sigma = 0$ case is replotted from Fig. 11b; also plotted are the frequency dependences of $\varepsilon''$ for $\sigma\tau_d/\varepsilon_0$ having various values relative to $\varepsilon_r - \varepsilon_u$. As $\sigma$ increases, it becomes increasingly difficult to discern the dipole loss peak. Roughly speaking, for $\sigma\tau_d/\varepsilon_0$ greater than about three times $\varepsilon_r$, the observed $\varepsilon''$ is entirely dominated by $\sigma$. (Ideally, even when $\sigma$ dominates the dipolar contribution to $\varepsilon''$, it should still be possible to observe the dipolar contribution to $\varepsilon'$; however, when $\sigma$ is large, electrode polarization effects tend to dominate the $\varepsilon'$ measurement as well. See Sec. 3.2.1).

A convenient way of displaying the $\varepsilon'$ and $\varepsilon''$ frequency dependences is in a Cole-Cole plot [29], where $\varepsilon''$ is plotted against $\varepsilon'$ with $\omega$ as a parameter. Figure 13 shows the Cole-Cole diagrams for the idealized cases illustrated in Figs. 11 and 12.

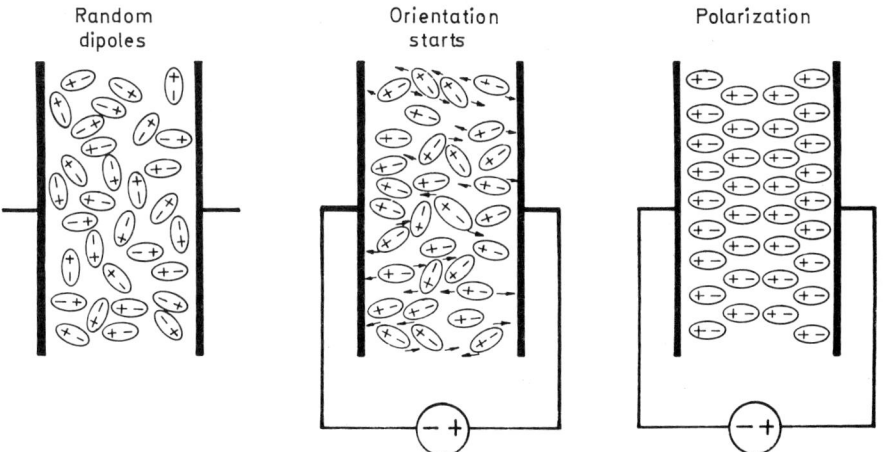

**Fig. 10.** Schematic illustration of dipole orientation process in the presence of an applied electric field. (Reprinted from Ref. [38] with permission of Gordon and Breach Science Publishers)

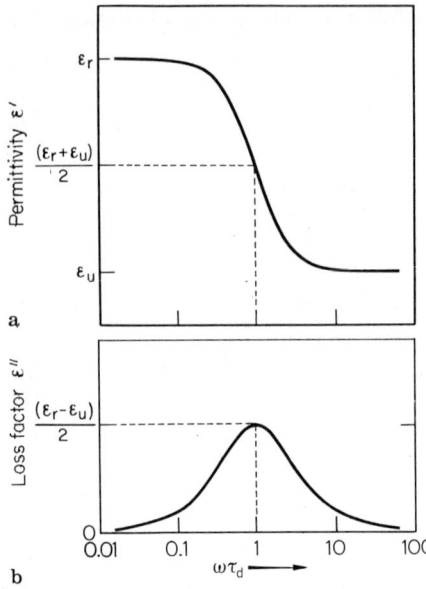

**Fig. 11a and b.** Debye single relaxation time model for dipole orientation showing **a)** permittivity and **b)** loss factor as a function of the product of the angular frequency ω and the dipole realxation time $\tau_d$. The relaxed permittivity is $\varepsilon_r$ and the unrelaxed permittivity is $\varepsilon_u$

Note that when $\sigma = 0$, the Cole-Cole diagram is a perfect semicircle, with endpoints at $\varepsilon_u$ and $\varepsilon_r$ and with a maximum $\varepsilon''$ value of $(\varepsilon_r - \varepsilon_u)/2$. As $\sigma$ increases, however, the Cole-Cole diagram approaches a vertical line with an intercept on the $\varepsilon'$ axis of $\varepsilon_u$.

In practice, the Cole-Cole diagrams observed differ from those of Fig. 13 in two ways. The first is due to electrode polarization, discussed in Section 3.2.1; the second is due to the fact that in real materials, the dipolar hindering mechanisms are

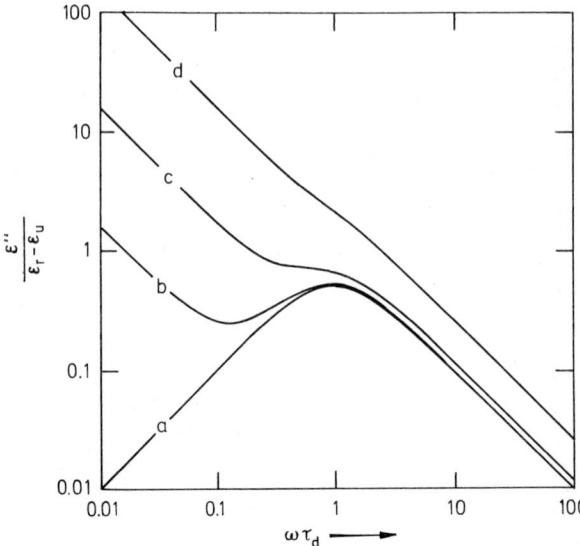

**Fig. 12.** Illustration of the effect of conductivity σ on the frequency dependence of the loss factor. a) $\sigma = 0$; b), c), d) $2\pi\sigma\tau_d/[(\varepsilon_r - \varepsilon_u)\varepsilon_0] = 0.1, 1, 10$

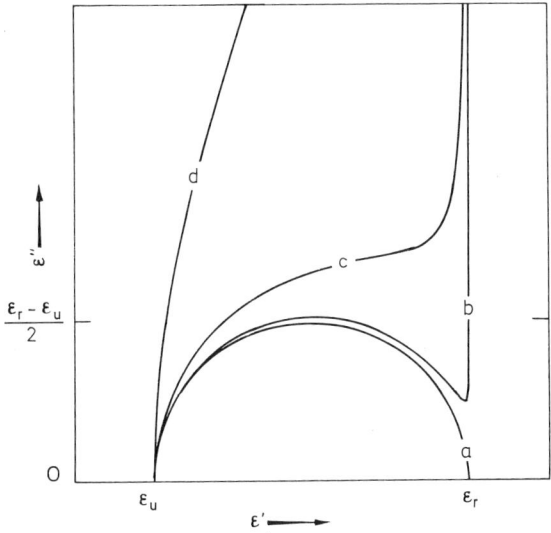

Fig. 13. Cole-Cole diagram illustrating effect of varying conductivity levels as in Fig. 12. Permittivity is from Fig. 11a

not characterized simply by one $\tau_d$, but by some distribution of relaxation times. A variety of empirical expressions have been proposed to describe real dipolar behavior; those by Cole and Cole [29], Davidson and Cole [30], and Williams and Watts [31,32] are most often cited. Although these expressions are empirical, each implicitly specifies a shape for the relaxation time distribution. A parameter β, known as the distribution parameter, characterizes the breadth of the distribution. This parameter ranges from 0 to 1, and each expression reduces to the Debye single relaxation time when β = 1. The Cole-Cole expression leads to a symmetric distribution of relaxation times about the mean value. The other two expressions lead to asymmetric distributions. The Williams-Watts formula assumes a polarization decay function of the form $\exp[-(t/\tau_d)^\beta]$, in constrast to the a polarization decay function of the form $\exp(-t/\tau_d)$ that would result from a single relaxation time.

Figure 14 compares the Cole-Cole diagrams for a single relaxation time (β = 1), the Cole-Cole expression with β = 0.5, and the Williams-Watts expression with

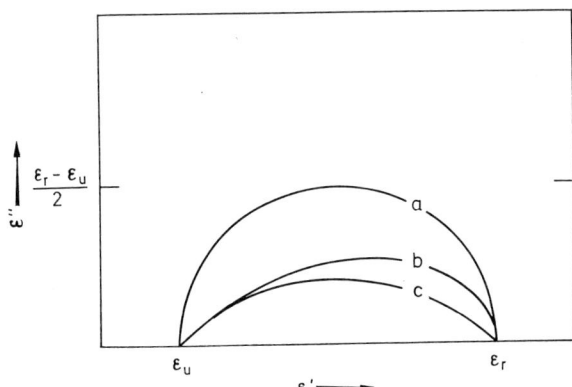

Fig. 14. Cole-Cole diagrams illustrating dipole relaxation behavior. a) Debye single relaxation time model. b) Williams-Watts expression with β = 0.5. c) Cole-Cole expression with β = 0.5

$\beta = 0.5$. The Cole-Cole function forms a symmetric arc, which approaches the intercepts with finite slope and has a maximum $\varepsilon''$ value less than $(\varepsilon_r - \varepsilon_u)/2$. The Williams-Watts function also forms a flattened arc, but is asymmetric. The shape of the Davidson-Cole function is very similar to the Williams-Watts function, as discussed by Lindsey and Patterson [33]. The evaluation of the Williams-Watts function requires numerical methods [33, 34]. Computer programs implementing this function from published tabular values are readily available [35].

## 3.2 Interface Effects

### 3.2.1 Electrode Polarization

The electrode-resin interface is normally electrochemically blocked at the low voltages used for dielectric measurements. This means that, just as in the case of an aqueous electrolyte, the applied electric field can polarize the electrodes by causing the accumulation of ion layers. The possibility of this effect was noted by Kienle and and Race [1], and has been modeled by Johnson and Cole [36]. The first quantitative analysis of electrode polarization effects in epoxy resin cure was reported by Adamec in 1972 [37]. The implications of Adamec's work, however, which are substantial, have been largely overlooked. New attention on this problem has resulted from the relative prominence of electrode polarization effects in low-frequency microdielectrometry data [38], but as discussed in Section 3.2.2, the effect is equally important in parallel-plate studies using blocking or release layers.

Figure 15 illustrates the effect. In direct analogy to the dipole orientation example of Section 3.1.2, the initially random distribution of ions becomes polarized in the electric field, with positive ions moving toward the negative electrode, and negative ions toward the positive electrode. Because the electrodes

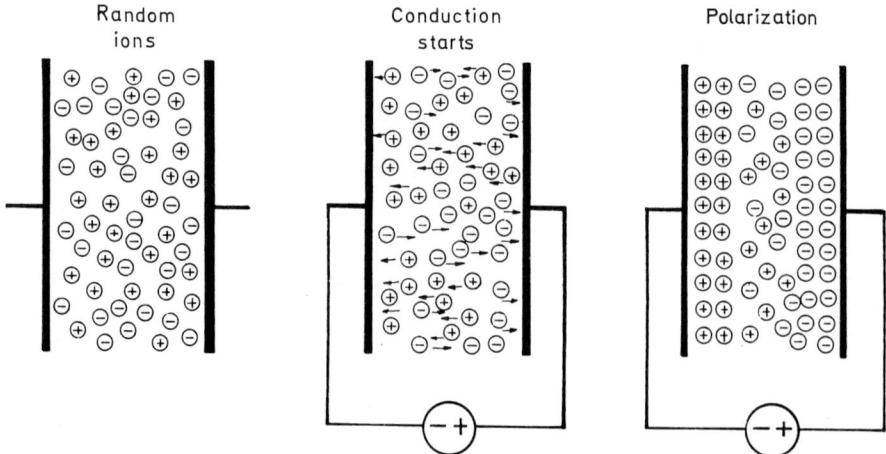

**Fig. 15.** Schematic illustration of ion conduction in the presence of an applied electric field leading to electrode polarization. (Reprinted from Ref. [38] with permission of Gordon and Breach Science Publishers)

are blocking, these ions accumulate at the electrodes, producing charged layers at both electrode. These are similar to the charge layers established by dipole orientation, but they can have a much greater charge per unit area. Thus, viewed from the electrodes, the measured sample capacitance $C_x$ can be much greater than that produced by the dipoles. The parallel $R_x - C_x$ circuit of Fig. 3 together with the "homogenous medium" assumptions implicit in the calibration used for the measurement result in an *apparent* permittivity $\varepsilon'_x$ which is much greater than the actual bulk permittivity $\varepsilon'$. The extent to which $\varepsilon'$ and $\varepsilon'_x$ differ depends on the sample inhomogeneity, as measured by the thickness of the charge layer relative to the inter-electrode distance.

To estimate the effects of electrode polarization, the equivalent circuit of Fig. 16 can be used. It shows a blocking layer capacitance $C_b$ (actually the series combination of two identical capacitors — one at each electrode interface) together with a parallel $R - C$ circuit representing the bulk material. The separate thicknesses of the blocking layer $2t_b$ and the total specimen length, L, must be used to construct the capacitances and resistance. The blocking layer capacitance $C_b$ has the value

$$C_b = \frac{\varepsilon' \varepsilon_0 A}{2t_b} \tag{3-8}$$

where, for simplicity, the permittivity in the blocking layer has been assumed to be the same as in the bulk specimen.

Qualitatively, it is apparent that if the conductivity is low, hence making R large, the sample will behave as a simple dielectric with permittivity $\varepsilon'$. However, when the conductivity is large enough so that the admittance $1/R$ becomes greater than $\omega C$, i.e., for $\tan \delta > 1$, then charging of $C_b$ through R becomes the dominant behavior of the circuit. Under these circumstances, the relative magnitudes of R and $C_b$ must be examined. If $\omega C_b > 1/R$, then charging of C, (i.e., electrode polarization) is not significant. However, when $1/R > \omega C_b$, the charging of $C_b$ becomes important. These arguments lead to two inequalities that must be satisfied if blocking layer effects are to be observed:

$$\tan \delta > 1 \tag{3-9}$$

and

$$\tan \delta > \frac{L}{2t_b} - 1 \tag{3-10}$$

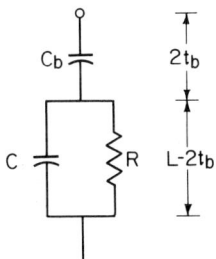

**Fig. 16.** Equivalent circuit for the electrode admittance in the presence of electrode polarization. (Reprinted from Ref. [38] with permission of Gordon and Breach Science Publishers)

The circuit model of Fig. 16 can be analyzed by transforming the circuit to the equivalent form of Fig. 3 and applying Eqs. (2-15) and (2-16). The resulting values of $\varepsilon'_x$ and $\varepsilon''_x$ depend on the properties of the medium ($\varepsilon'$, $\varepsilon''$, and tan $\delta$) *and* on the ratio $L/2t_b$, as follows [38]:

$$\varepsilon'_x = \varepsilon' \frac{L}{2t_b} \left[ \frac{(\tan \delta)^2 + \left(\frac{L}{2t_b}\right)}{(\tan \delta)^2 + \left(\frac{L}{2t_b}\right)^2} \right] \tag{3-11}$$

$$\varepsilon''_x = \varepsilon'' \frac{L}{2t_b} \left[ \frac{\left(\frac{L}{2t_b} - 1\right)}{(\tan \delta)^2 + \left(\frac{L}{2t_b}\right)^2} \right] \tag{3-12}$$

The experimental loss tangent, tan $\delta_x$, is obtained from $\varepsilon''_x/\varepsilon'_x$. In the ideal case of no electrode polarization ($L \gg t_b$), these Equations reduce to $\varepsilon'_x = \varepsilon'$, $\varepsilon''_x = \varepsilon''$, and tan $\delta_x$ = tan $\delta$, as expected.

The experimental Cole-Cole diagrams in the presence of a polarization layer, even for a medium in which $\varepsilon'_d$ is zero, resemble Debye-model semicircles. This is characteristic of systems with a single relaxation time, which in this case is the time required to charge the blocking layer through the medium. Figure 17 shows experimental data for a DGEBA resin on a microdielectrometer sensor [38] superimposed on Cole-Cole diagrams calculated from Eqs. (3-11) and (3-12). The temperature and frequency were varied to achieve a wide range of bulk $\varepsilon''$ values, and for two of the curves, polyimide coatings of 1500 Å and 1.2 µm were

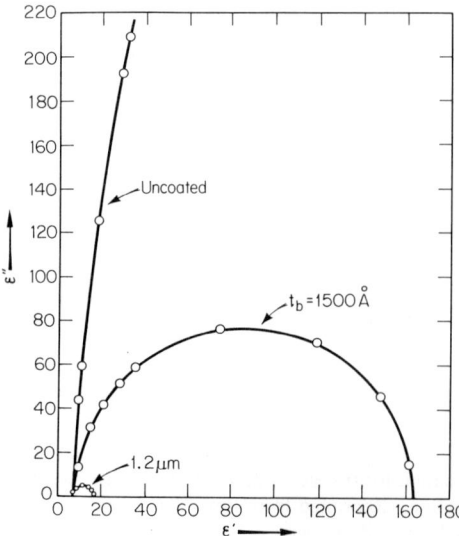

Fig. 17. Cole-Cole diagram of the dielectric properties of a DGEBA epoxy resin measured on a microdielectrometer sensor with blocking layers of indicated thickness. (Reprinted from Ref. [38] with permission of Gordon and Breach Science Publishers)

used to cover the electrodes. Similar Cole-Cole plots have been reported by Zukas, et al.,[39] in a parallel-plate epoxy-cure experiment in which a PTFE blocking film was used on one electrode. Even in the absence of these added coatings, nominally bare electrodes show blocking characteristics, with a blocking layer thickness of about 60 Å at the interface between epoxy resins and aluminium electrodes [38].

The presence of electrode polarization layers can have a profound effect on the interpretation of dielectric cure data. This is discussed in the following Section.

### 3.2.2 Blocking and/or Release Layers

It is common practice, either to bring the sample $\tan \delta_x$ within the range of a particular instrument, or for convenience in removing electrodes from the sample, to insert a thin blocking layer or release layer between the plates of a parallel-plate capacitor, or across the surface of a comb electrode. In both cases, the added blocking layer thickness completely changes the characteristic of the observed cure data.

In a typical ramped cure, the specimen begins with a relatively small loss factor, which initially increases as the temperature is increased due to a decrease in viscosity, but which later decreases due to the effects of cure. This has been illustrated schematically in Fig. 18 [38], where the topmost curve represents the initial increase and subsequent decrease of the actual bulk $\tan \delta$ for a specimen during cure. The remaining curves are the values of $\tan \delta_x$ that would be observed experimentally in the presence of either an electrode polarization layer or an added blocking/release layer for various values of the ratio $L/2t_b$. Note that even a thin release layer, for example a 25 µm (.001") layer placed between 5 mm spaced plates, results in an $L/2t_b$ ratio of 100. Examination of Fig. 18 shows that for this value of $L/2t_b$, the *maximum* value of $\tan \delta$ actually produces a *minimum* in the experimental $\tan \delta_x$, and that two subsidiary maxima appear in $\tan \delta_x$.

The literature is rich in examples of this "upside-down" minimum and the secondary maxima arising from electrode polarization [40-43]. Figure 19 shows Lawless' superposition of a temperature ramp, the viscosity, and the measured $\tan \delta_x$ for the cure of an Avco 5505 epoxy resin with parallel plates in the presence

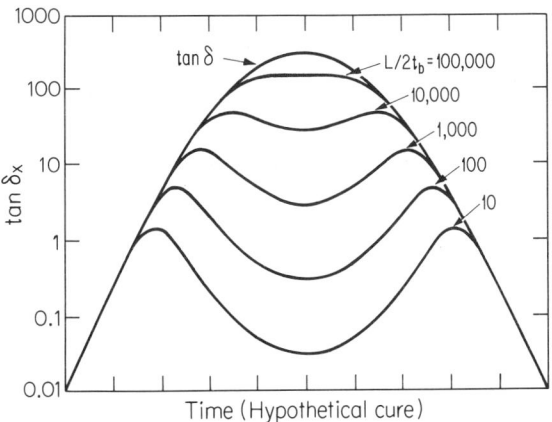

Fig. 18. Schematic illustration of loss tangent behavior during a hypothetical cure for indicated values of the ratio $L/2t_b$. (Reprinted from Ref. [38] with permission of Gordon and Breach Science Publishers)

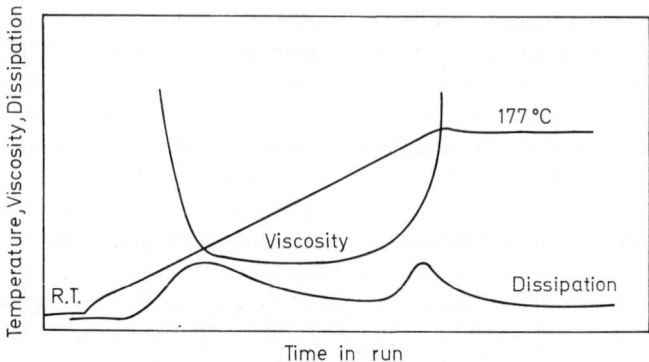

**Fig. 19.** Plot of viscosity and dissipation factor during cure of an epoxy resin illustrating "camelback" in dissipation factor and the correspondance of the minimum viscosity with the minimum between the two dissipation factor peaks. (Reprinted from Ref. [42] with permission of the Society of Plastics Engineers)

of Kapton release layers [42]. The minimum viscosity corresponds to the minimum in tan $\delta_x$, but as has been emphasized, this tan $\delta_x$ minimum actually corresponds to the maximum intrinsic tan $\delta$. The agreement of the tan $\delta$ maximum with the viscosity minimum is a reasonable result since the viscosity minimum occurs well before gelation, where the resistivity and viscosity are correlated. The two subsidiary maxima have been variously ascribed to "events" such as "flow" or "gelation", where in fact, they simply identify the point in the experiment at which polarization of the blocking layer becomes the dominant electrical feature of the sample.

The dependence of results on $L/2t_b$ also explains why parallel-plate experiments can have calibration and reproducibility problems if the plate spacing is not rigorously controlled during a cure. Comb electrodes, in spite of the complexity of their basic calibration, offer the distinct advantage of having a rigid and reproducible electrode geometry, which permits quantitative evaluation of blocking layer effects.

### 3.2.3 Fibers and Fillers

There are several possible effects of internal inhomogeneities such as fibers and fillers on the measured dielectric properties of composites. Interfacial polarization can be established at any interface between media of different conductivities, leading to semicircular Cole-Cole diagrams as discussed in Section 3.2.1. The parameters of the Cole-Cole diagram depend both on the dielectric properties of the media and on their geometry. The appearance of these apparent Debye-like dielectric relaxations based on nonuniform conductivity within the sample is generally referred to as the Maxwell-Wagner effect, and is well established in the dielectrics literature [23].

For non-conducting fibers, such as glass, the matrix resin is the more conductive phase, at least early in cure, and one would expect some internal polarization effects to be visible in parallel-plate data. However, in spite of a large body of literature on glass fiber composites (see Sect. 5), we have found no clearly documented cases of Maxwell-Wagner effects in fiber-reinforced composites. We speculate that the widespread practice of using release films may obscure the effects of internal polarizations. Another possible explanation is the difficulty of obtaining quantitative and reproducible comparisons between neat resin and matrix resin data. For example, Bidstrup, et al., [44], using microdielectrometry, have noted a quantitative discrepancy between the dielectric properties of (1) a matrix resin measured in-situ during cure of a glass circuit-board laminate and (2) a sample of neat resin flaked from the staged prepreg and cured apart from the glass fibers. The in-situ resin showed higher conductivity, possibly attributable to additional ions from the fiber or fiber sizing; no definitive explanation has been established.

In the case of graphite fibers, the fiber itself is more conductive than the resin, which can significantly affect parallel-plate measurements. Nevertheless, Pike, et al., in 1971 [45] reported successful dielectric monitoring of cure in graphite-epoxy and graphite-polyimide laminates using parallel-plate copper electrodes separated from the laminate by glass release cloth. Their data show the pair of maxima cited in the previous section, which may be due either to internal interfacial polarization or to electrode polarization. As with the non-conducting fibers, there has been no definitive observation of Maxwell-Wagner effects in graphite composites.

## 4 Effects of Temperature and Cure

### 4.1 General Issues

This Section addresses the effects of temperature and cure on the various microscopic mechanisms presented in Section 3. Since temperature is a key variable in determining the rate of cure processes, and since all of the microscopic mechanisms are dependent directly on temperature as well as indirectly on time and temperature through the cure reaction, it will prove useful to examine, first, the temperature behavior of epoxy resins themselves (without curing agent), and, second, the corresponding behavior in systems undergoing cure. Of the various epoxy resins, most of the quantitative studies involve the diglycidyl ether of bisphenol-A (DGEBA), having the general formula

$$H_2C-CH-CH_2-O-\left[\phantom{x}\bigcirc-\underset{CH_3}{\overset{CH_3}{C}}-\bigcirc-O-CH_2-\underset{OH}{\overset{}{CH}}-CH_2-O\right]_n \bigcirc-\underset{CH_3}{\overset{CH_3}{C}}-\bigcirc-O-CH_2-HC-CH_2$$

where n indicates the degree of chain extension of the resin.

Figure 20 shows the temperature dependence of the permittivity and loss factor for a DGEBA epoxy resin (EPON 828; $n \cong 0.2$) in the vicinity of its glass transition ($T_g = -17\ °C$), measured at frequencies between 0.1 and 10,000 Hz [46]. At temperatures well below $T_g$, the permittivity at all frequencies has a value of of 4.2 (the unrelaxed permittivity), and the loss factor is below 0.1. As the temperature approaches $T_g$, the dipoles gain sufficient mobility to contribute to the permittivity, with evidence of this mobility increase occurring first at the lowest frequency. With a further increase in temperature, the permittivity for a given frequency levels off at the relaxed permittivity, which then decreases due to increasing temperature

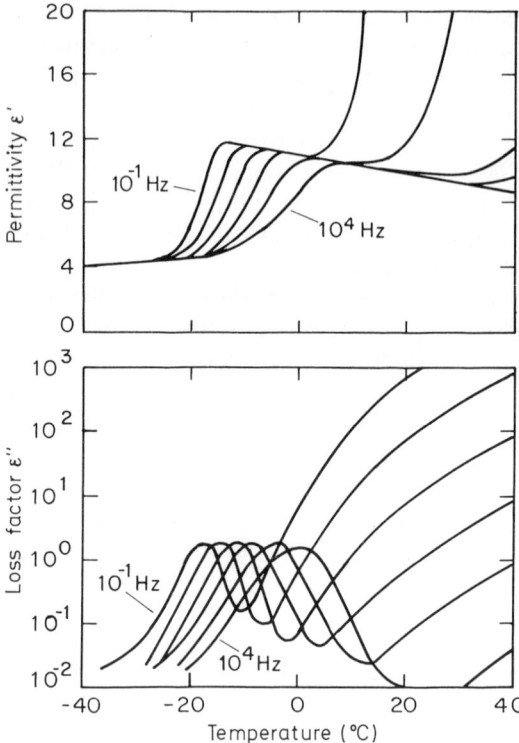

**Fig. 20.** Plot of permittivity and loss factor versus temperature for EPON 828 resin in the vicinity of the glass transition. (Reprinted from Ref. [46] with permission of the authors)

(see Sect. 4.3), and then abruptly increases again as a result of electrode polarization (Sect. 3.2). At each frequency, a dipole peak is observed in the loss factor, which then rises continuously with temperature due to an increasing ionic conductivity. The frequency at which the dipole loss peak occurs is proportional to the average dipole mobility. The ionic conductivity is proportional to the ionic mobility. Both the frequency of maximum loss and the ionic conductivity increase by many orders of magnitude over a narrow temperature range, a characteristic of relaxation processes very close to the glass transition temperature.

Figure 21 shows the permittivity and loss factor for an isothermal cure (137 °C) of DGEBA (EPON 825; n $\cong$ 0) with diaminodiphenyl sulfone (DDS)[47]. To a first approximation, the data are the mirror image of Fig. 20, supporting the idea that the effect on electrical properties of the increase in $T_g$ during isothermal cure might be similar to the effect of a decrease in temperature at fixed $T_g$. Examination of Fig. 21 shows one important difference between the temperature and cure dependences, namely that the relaxed permittivity *decreases* with cure time under isothermal conditions. This is a direct result of the changing chemistry, as discussed further in Section 4.3. The detailed behavior of the dipolar mobility is examined in Section 4.4, and of the ionic mobility in Section 4.5.

We shall see that throughout the literature there has been an implicit assumption of thermally activated processes, both for cure kinetics *and* for the intrinsic dipolar and ionic mobilities. However, it is well known that reaction kinetics become diffusion controlled at the later stages of cure, which leads to deviations from simple rate

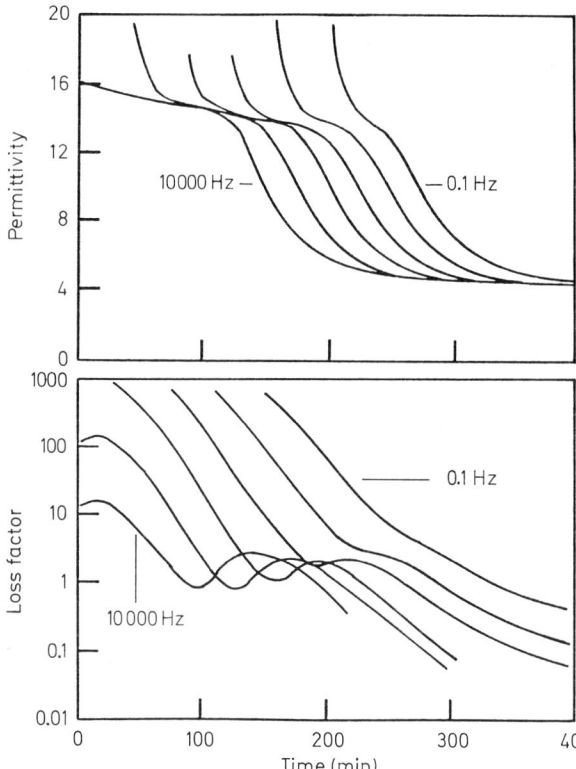

Fig. 21. Plot of permittivity and loss factor versus time during the cure of a low molecular weight DGEBA resin (EPON 825) with DDS at 137 °C. (Reprinted from Ref.[47] with permission of the Society for the Advancement of Material and Process Engineering)

equations toward kinetics dominated by diffusive mobilities[48]. Furthermore, post-cure studies of dipolar mobilities[49-52] and studies of ion mobilities in epoxy resins[46] suggest that their temperature dependences are described by the Williams-Landel-Ferry equation[53] rather than by Arrhenius behavior. Taken together, it is fair to say that a comprehensive model of dielectric properties during a thermoset cure, including proper chemical kinetics and correct relations between state of cure, temperature, and the various dielectric parameters, has not yet been developed. However, there have been significant strides toward this goal, as the remainder of this article will demonstrate.

Two more difficulties require comment. The first is that in most of the early literature, authors did not recognize the importance of electrode polarization, and, hence, failed to make quantitative allowance for the presence of blocking and/or release layers. Thus, in most cases, it is not possible to reconstruct quantitative bulk properties from the data presented. (The present authors were not immune. They reported a correlation between a "dielectric relaxation time" and viscosity[54], failing at that time to realize that the relaxation time being studied was actually the characteristic time for electrode polarization, and, hence was dominated by conductivity.)

A second difficulty is due to an accident. In Delmonte's 1959 study of several epoxy resins[55], the time to gel and the time to vitrify were very close to one another. Delmonte identified the dipole peak as arising from the gelation event, a conclusion which has been widely cited. The incorrectness of this assignment was noted as early as 1965 by Olyphant [56], who in an excellent tutorial review of dielectric properties of curing systems, clearly identified the dipole peak with vitrification. Olyphant also stated, in agreement with the pioneering work of Warfield and Petree [57, 58], that there are no electric "events" accompanying gelation. In spite of these unambiguous statements in the early literature, attempts at the electrical identification of gelation continues as a theme, aggravated by the artifacts associated with conductivity and electrode polarization. This accidental assignment of electrical signatures to gelation is discussed further in Sect. 4.5.

## 4.2 Relation with Chemical Kinetics

In order to understand the relation between the chemical kinetics of a thermosetting system and the corresponding changes in dielectric properties, it is important to separate the various mechanisms giving rise to such changes. The relaxed permittivity is directly sensitive to the changing chemical composition, because it depends on the concentrations of the various polar molecular segments (see Sect. 4.3.). The dipolar mobilities (Section 4.4) and the ionic mobilities (Sect. 4.5) depend on the extent of reaction primarily through the change in $T_g$ during cure, and this dependence on $T_g$ can be described by the WLF Equation. The recognition of this WLF dependence is so recent that there have been no kinetic-dielectric studies which include it. Instead, a variety of empirical relations between dielectric properties and cure have been reported, principally in epoxy systems. These kinetic studies are reviewed in this Section.

The simplest form of empirical kinetic information comes from rates of change of experimental properties, such as the time-rate-of-change of conductivity, or the cure time to reach a specific event, such as a dipole loss peak. The earliest quantitative analysis of such data was by Kienle and Race [1] who established a linear correlation between extent of esterification of alkyd resins and the corresponding log of resistivity (the resistivity, $\varrho$, is the inverse of conductuctivity). Practical examples of the use of such empirical kinetic information are reviewed in Section 5. Warfield and Petree [57, 58], working both with diallyl phthalate and epoxy-amine resins, studied the relation between the log of resistivity and cure time at various temperatures. By identifying $d \log (\varrho)/dt$ as an empirical reaction rate (an unjustified assumption; see Sect. 4.5), they inferred a temperature dependence for the reaction rate, and calculated an activation energy.

The first attempt to relate changing dielectric properties to kinetic rate equations was by Kagan et al. [59], working with a series of anyhydride-cured epoxies. Building on Warfield's assumed correlation between $d \log (\varrho)/dt$ and $d\alpha/dt$, where $\alpha$ is the extent of epoxide conversion, they assumed a proportionality between $\alpha$ and $\log (\varrho)$, and modeled the reaction kinetics using the equation

$$d\alpha/dt = k(1 - \alpha)^m \qquad (4\text{-}1)$$

where k is the rate constant and m is the empirical reaction order. Similar analyses using mth-order kinetics have been done by Acitelli et al. [60], working with diglycidyl ether of Bisphenol-A cured with m-phenylenediamine, and by Adamec [37], working with a commercial epoxy novolac resin (Dow DEN 438) cured isothermally with 3phr $BF_3$-monoethylamine. Kagan extracted empirical rate constants and activation energies from his resistivity data, but did not compare these results with other methods. Actelli used both DSC and IR data prior to gelation to compare against resistivity data obtained after the break in log ($\varrho$) versus time (which they claim is gelation; see Section 4.5 for a discussion). They find the same rate constant from all methods, but the empirical reaction order they must use varies enormously depending on the experimental method and on the temperature range. A similar problem in reaction order was reported by Adamec.

One difficulty with such empirical approaches based on the dielectric properties is that they tend to oversimplify the chemical kinetics, which usually are obtained from independent thermal analysis studies coupled with other measurements [61]. The epoxy-amine reaction, for example, is catalyzed by hydroxyl, hence becomes autocatalyzed as the cure proceeds [62]. In addition, an etherification reaction can compete with the amine reaction, leading to complex branched kinetics. Sourour and Kamal [63], working with DGEBA cured with m-phenylenediamine, developed an autocatalyzed kinetic model for epoxy-amine systems using isothermal DSC measurements. Huguenin and Klein have extended the kinetic studies to the diffusion-controlled regime [48]. The use of these models to analyze the change in relaxed permittivity during the amine cure of a difunctional DGEBA epoxy is discussed in the following section.

In other chemical systems, such as tetrafunctional epoxies, polyimides, phenolics, and polyesters, there have been few attempts [64, 65] to establish quantitative relationships between chemical kinetics and dielectric properties.

## 4.3 Relaxed Permittivity

The relaxed permittivity $\varepsilon_r$ measures the maximum dipolar alignment that can be achieved at a given temperature and chemical state. It is observed at frequencies sufficiently low to allow dipolar alignment, but sufficiently high to avoid electrode polarization effects. In Figs. 20 and 21, the variation of $\varepsilon_r$ with either temperature or cure must be obtained by piecing together the small intervals at each frequency where the permittivity data follow $\varepsilon_r$. Generally, $\varepsilon_r$ decreases with increasing temperature, and for the case of epoxy-amine cure, decreases with increasing cure time.

The quantity $\varepsilon_r$ is directly sensitive to the detailed chemical composition of the sample. However, the quantitative theory that relates the observed $\varepsilon_r$ to the concentrations and dipole moments of the various polar segments present has proved quite difficult to use. The simplest approach is based on the Clausius-Mosotti equation as modified for permanent moments by Debye [28]. The Debye approach, although overly simple, revealed that $\varepsilon_r$ should decrease with increasing temperature, and should reflect changing concentrations of polar constituents during a reaction.

The first attempt to use these ideas in epoxy cure was by Fisch and Hofmann [66], but their assignment of permittivity changes to changes in polar group concentrations was marred by what we interpret as electrode polarization effects. Blyakhman et al. [51, 52], examined the post-cure dielectric permittivity and loss tangent of anhydride-cured and amine-cured DGEBA resins. Their data are similar to that for the EPON 828 resin in Fig. 20. Based on an observed variation of the post-cure $\varepsilon_r$ with the molecular weight of the DGEBA resin, they argue that the principal contribution to $\varepsilon_r$ is from hydroxyether groups in the epoxy chain. Their conclusion is reasonable, since their experiments were after cure, and the highly polar curing agent and epoxides had reacted.

Huraux and co-workers, in a series of papers [67-70], and more recently, Sheppard [71], have attempted quantitative interpretation of $\varepsilon_r$ during an epoxy cure in terms of the changing concentrations of constituent polar groups. They used a theory by Onsager which improves Debye's original theory to account for local dipole fields [see, for example, Ref. [23]]. The Onsager theory, expressed below, requires some explanation:

$$\frac{(\varepsilon_r - \varepsilon_u)(2\varepsilon_r + \varepsilon_u)}{\varepsilon_r(\varepsilon_u + 2)^2} = \sum_i \frac{N_i(\mu_i)^2}{9\varepsilon_0 kT} \qquad (4\text{-}2)$$

The Equation assumes that all dipoles are independent. It relates the values of $\varepsilon_r$ and $\varepsilon_u$ measured at temperature T to the concentrations N and dipole moments $\mu_i$ of the polar species (k is Boltzmann's constant and $\varepsilon_0$ is the permitivity of free space).

Huraux et al., [68] used the Onsager theory to analyze the permittivity of a DGEBA resin (n = 0.2) cured with a cycloaliphatic diamine. They assumed that all of the polar behavior could be attributed to epoxide groups (see below for an improved assumption), and thus extracted the time rate of change of epoxide groups from the measured rate of change of $\varepsilon_r$ during cure using the time derivative of both sides of Eq. (4-2) ($\varepsilon_u$ did not change either with temperature or cure). The $\varepsilon_u$ and $\varepsilon_r$ values were obtained by examining the Cole-Cole plots of the dipole relaxation data and extrapolating the arcs to their endpoints [67] (see Section 4.4 for additional details). The rate of epoxy consumption was integrated to obtain the extent of conversion, which was then to calculate an average molecular weight which was found to correlate with viscosity through a power law.

Soualmia et al. [70] extended this approach to a specific rate equation for the chemical kinetics, which they followed independently by titration for unreacted epoxide. The system studied was a low molecular weight DGEBA resin cured with a cycloaliphatic diamine. All their measurements were at room temperature. The titration data agreed well with a second-order rate equation for the disappearance of epoxide, but in attempting to assign the reaction-dependent decrease in the dielectrically derived $\Sigma N_i(\mu_i)^2$ entirely to the disappearance of epoxide groups (i.e., to decreases in $N_i$ for epoxides), they were forced to the unsatisfactory conclusion that the effective epoxide moment was changing during cure. Their dipole moment values were $5.3 \times 10^{-30}$ C-m per epoxide group for the pure resin, increasing to $7.1 \times 10^{-30}$ C-m when initially mixed with during agent, and subsequently decreasing to $6.24 \times 10^{-24}$ C-m at the end of cure.

One implication of Soualmia's result is that the curing agent plays a significant role in determining $\varepsilon_r$. This means that both the curing agent concentration and moment must be determined as part of the analysis. Sheppard [71] has attempted this

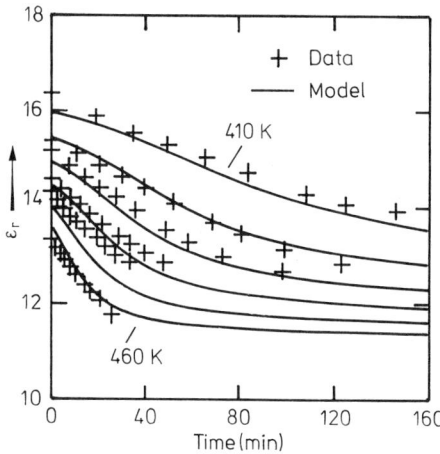

**Fig. 22.** Plot of relaxed permittivity versus time for a low molecular weight DGEBA resin (EPON 825) cured isothermally with DDS at temperatures between 410 K and 460 K. Crosses represent experimental data; the solid curve represents the model described in the text. (Reprinted from Ref. [71] with permission of the Society of Plastics Engineers)

for the DGEBA (n = 0) system cured with DDS. Figure 22 shows $\varepsilon_r$ versus cure time at several temperatures, obtained from data like that of Fig. 21. The extent of conversion was determined as a function of time at each temperature from independent DSC data analyzed with the autocatalyzed reaction kinetics model of Sourour and Kamal [63], which assumes that the primary and secondary amines react at the same rate, and further assumes that homopolymerization of the epoxy is negligible. This model fits the DSC data up to about 60% epoxy conversion, at which point the reaction kinetics become noticable diffusion controlled. Up to 60% conversion, however, the kinetic model can be used to calculate the concentrations of primary, secondary, and tertiary amines as functions of time at each cure temperature. Based on the kinetic model, Sheppard extracts the temperature dependence of $\varepsilon_r$ *at constant conversion*. Figure 23 shows the corresponding Onsager theory plot, i.e., the left-hand-side of Eq. 4-2 plotted against 1/T. Lines extrapolated through the data points do not intersect the origin, indicating that the simple Onsager theory fails.

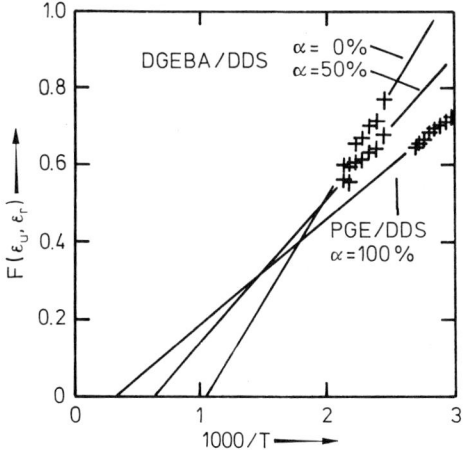

**Fig. 23.** Test plot of the Onsager relation (Eq. (4-2)) for the data of Figure 22 at different extents of conversion as determined by DSC measurements. (Reprinted from Ref. [71] with permission of the Society of Plastics Engineers)

A similar failure of the Onsager model has been reported for polyacetaldehyde by Williams [72], and in epoxy resins by Sheppard [73]. Williams was able to analyze the polyacetaldehyde data in terms of a temperature dependent correlation between neighboring dipoles based on a conformational model of the polymer chain, and Sheppard has noted [35] that qualitatively similar conformational issues should apply to the epoxide and hydroxyl groups in the epoxy.

Both the slope and intercept point in Fig. 23 change with increasing extent of conversion, which may be due to chainging $N_i'$s and a changing interdipole correlation. To simplify the analysis, an empirical modification to the Onsager equation was used to analyze the data:

$$\frac{(\varepsilon_r - \varepsilon_u)(2\varepsilon_r + \varepsilon_u)}{\varepsilon_r(\varepsilon_u + 2)^2} = \sum_i \frac{N_i(\mu_i)^2}{9\varepsilon_0 k}\left[\frac{1}{T} - \frac{1}{T_\alpha}\right] \quad (4\text{-}3)$$

where $T_\alpha$ is the empirical temperature at which the plot intersects the axis for extent-of-conversion $\alpha$. It is implicitly assumed that the concentrations of each polar group, $N_i$, depend on $\alpha$ based on the chemical kinetics. The data in Fig. 23 were analyzed using this model assuming three polar groups, unreacted epoxide, primary amines, and reacted amines (both secondary and tertiary). Except for the dipole moments for each species, $\mu_i$, all quantities in Eq. (4-3) were known up to 60% conversion. A fit to the data yielded the agreement shown as the solid lines in Fig. 22 using a dipole moment of $7.6 \times 10^{-30}$ C-m for the epoxide, $14.8 \times 10^{-30}$ for the primary amine, and $12.6 \times 10^{-30}$ C-m for the reacted amine. The larger moment values for the amine constituents demonstrate the significant contribution they make to the total $\varepsilon_r$ value.

The work of Huraux, Soualmia, and Sheppard has demonstrated that it is possible to make a quantitative interpretation of the $\varepsilon_r$ values (within the limits of an admittedly difficult theory), and that the $\varepsilon_r$ value is directly linked to the chemical changes during cure. However, because of the complications introduced by dipole correlations, the relaxed permittivity is not a useful tool for routine quantitative determination of polar reactive group concentrations during cure.

## 4.4 Dipolar Relaxations

This Section addresses the effects of temperature and cure on the dominant dipolar relaxation, i.e., the $\alpha$-transition between the unrelaxed and relaxed permittivity, with its associated loss-factor peak. As illustrated in Figs. 20 and 21, this dipolar relaxation is observed as the temperature increases through $T_g$, or during a cure, as $T_g$ increases toward and even through the cure temperature.

Discussion of the dipolar relaxation involves two issues: first, the average dipolar mobility at a given temperature and degree of conversion, as measured by the frequency of the maximum in the loss factor $f_{max}$ (or by its reciprocal, the typical dipolar relaxation time $\tau_d$), and, second, the detailed distribution of relaxation times as measured by the frequency dependence of the permittivity and loss factor. In spite of the clear evidence that the dipolar relaxation is associated with the glass transition

[56)], i.e. with *vitrification* of a curing resin, there is also evidence that early in cure, the dipolar relaxation time correlates with viscosity [67-70, 74, 75]. Because the viscosity becomes infinite at *gelation*, it is almost "natural" (yet incorrect) to assign the dipolar relaxation to gelation [55]. The conflict in the literature can be resolved by examining the fundamental role that polymer chain mobility plays in both the viscosity and dipolar mobility.

It is helpful to begin the discussion with dipolar relaxations in pure epoxy resins, where the complexity of the gelation-vitrification issue is absent. The WLF Equation [53], which is widely used to model the temperature dependence of mobility-related material properties, has the form

$$\log (a_T) = \pm \frac{C_1(T - T_S)}{C_2 + T - T_S} \qquad (4\text{-}4)$$

where the shift factor $a_T$ is defined as the ratio of the property of interest at temperature T to the same property at a reference temperature $T_S$ (taken as $T_g$ in the present discussion), and where the sign of the right-hand-side is chosen according to the sign of the temperature dependence of the property being analyzed. $C_1$ and $C_2$ are constants, initially speculated as having universal values, but actually depending on the material system and on the property being measured. The important point is the dependence of the shift factor on the *difference* $T - T_S$.

Sheppard [46] has measured the temperature dependence of $f_{max}$ for a homologous series of DGEBA epoxy resins with n ranging from 0 to 12. The results are plotted in Arrhenius form in Fig. 24. While it is tempting to assign activation energies to these data, careful analysis shows that there is curvature in the data, and that the results are better described by the WLF Equation. The solid curves in Fig. 24 are Sheppard's WLF fit to the data, which yielded constants $C_1$ and $C_2$ that are comparable to the "universal" $C_1$ and $C_2$ values. However, the $C_1$ constant for the high molecular weight resins (n > 2) was 50% of the value for the low MW samples (n = 0, 0.2). The $C_2$ value decreased only slightly with n, and had a mean value of 54 °C. The $C_1$ decrease was attributed to the increasing dipolar contribution of hydroxyether

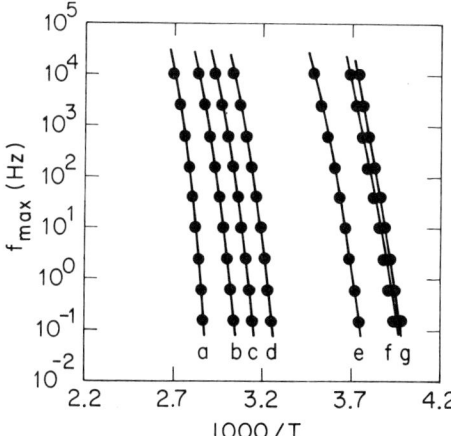

**Fig. 24.** Arrhenius plot of the frequency of maximum dipole loss for DGEBA epoxy resins of varying molecular weights. a) n = 12.1; b) 5.1; c) 3.4; d) 2.1; e) 0.6; f) 0.2; g) 0. (Reprinted from Ref. [46] with permission of the authors)

moieties in the higher molecular weight resins, and to the speculation that these require less free volume to relax than the polar epoxide endgroups.

To connect these results to the dielectric relaxation in *curing* systems, it is useful to examine the *post-cure* data of Shito et al. [49, 50] and of Blyakhman et al. [51, 52]. Figure 25 shows the fit of Shito's data for $\tau_d$ (the reciprocal of $f_{max}$) to the WLF Equation for a series of DGEBA resins cured with a variety of anhydrides [50]. In this case, the data were fit to the universal WLF Equation, using the reference temperature as an adjustable parameter, which fell in the range 55–61 °C above $T_g$ for the various samples.

The epoxy resin data and the post-cure data, taken together, show that the dipolar relaxation is associated with the temperature dependence of the polymer chain mobility in the vicinity of the glass transition. The WLF analysis of the dipolar relaxation *during cure* has not been carried out. In order to complete the analysis, correlated measurements of $T_g$, extent of cure, and dielectric properties must be made as functions of cure time and temperature. In the absence of such definitive studies, various indirect methods have been employed to analyze dielectric relaxations in curing systems, as described below.

The first method has been to draw on an expected correlation between the relaxation time and viscosity (prior to gelation). The correlation derives from elementary considerations of viscous drag on dipoles, and was originally predicted by Debye [28]. Huraux and co-workers were the first to report quantitative correlations between the mean dipole relaxation time and viscosity during epoxy cure [67–70]. Lane, et al., have also reported such a correlation [74, 75]. Furthermore, the viscosity has recently been analyzed from a WLF point of view. Tajima and Crozier [76], and Apicella, et al., [77] have demonstrated that the temperature dependence of the viscosity during an epoxy cure can be described by a WLF equation with a reference temperature that increases with increasing extent of conversion.

The implication of all these results is a prediction that the *temperature dependence of the dipole relaxation time at fixed conversion will obey the WLF Equation and*

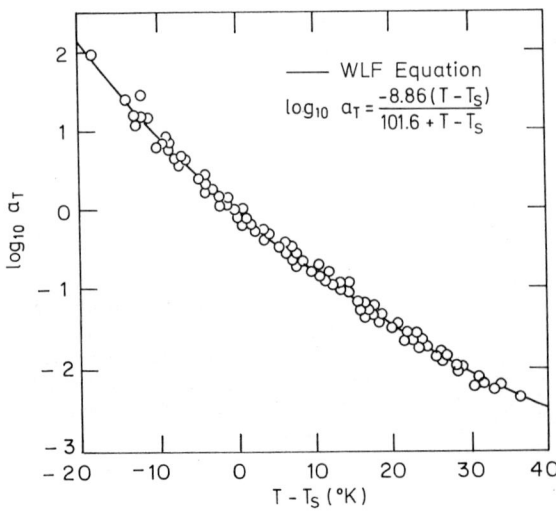

Fig. 25. Shito's test plot of the Williams-Landel-Ferry equation for the dipole relaxation time in an anhydride-cured epoxy. (Reprinted from Ref. [50] with permission of John Wiley and Sons, Inc.)

*that the reference temperature will track the glass transition as it increases during cure.* Work on this is currently under way at several laboratories.

A second method for analyzing the dipolar relaxation in curing systems is to compare isothermal cure time required to reach the dipole loss peak with cure time required to reach other events, such as vitrification as measured with torsional braid analysis. One interesting result of Sheppard's analysis of pure epoxies [46] is that the loss peak measured in the frequency range 1–3 Hz tracks $T_g$ as measured by DSC. Extrapolating from this result, one expects that during cure the time to reach the dipole loss peak at about 1 Hz should fall near the vitrification boundary of the time-temperature-transformation (TTT) diagram [78]. In most curing systems, however, the low frequency dipole loss peaks at typical cure temperatures are obscured by the conductivity; as a result, the dipole peak is observable only at higher frequencies. Nevertheless, both the present authors [47] and Zukas, et al. [79] have found that the temperature dependence of the time to reach the dipole peak agrees with the temperature dependence of the time to reach vitrification. Figure 26 illustrates the loss-peak data at two frequencies for EPON 825 cured with DDS superimposed on the TTT diagram measured by Enns and Gillham using torsional braid analysis [47]. While the parallel trend with cure temperature is clear, the absolute time to reach the dipole peak at higher frequencies is systematically shorter than the time to reach vitrification.

We now examine briefly the distribution of relaxation times. The most direct evidence for a distribution of relaxation times is the departure of the Cole-Cole plot from a perfect semicircle, as illustrated in Fig. 14. Whether one uses a Cole-Cole [29], Davidson-Cole [30] or a Williams-Watts [31, 32] function to describe this distribution, a decrease in the parameter β to less than unity indicates a distribution of relaxation times. As early as 1965, Olyphant [56] noted that β was less than unity, and decreased during the cure of an anhydride-cured epoxy, indicating a broader distribution of relaxation times after cure. Huraux and co-workers [67, 70] similarly report a broadening of relaxation time distribution during cure. Sheppard [35] has observed a considerable decrease in the Williams-Watts β with increasing molecular weight of DGEBA resins, and also during the cure of EPON 825 with DDS. Daly and Pethrick [80] examined

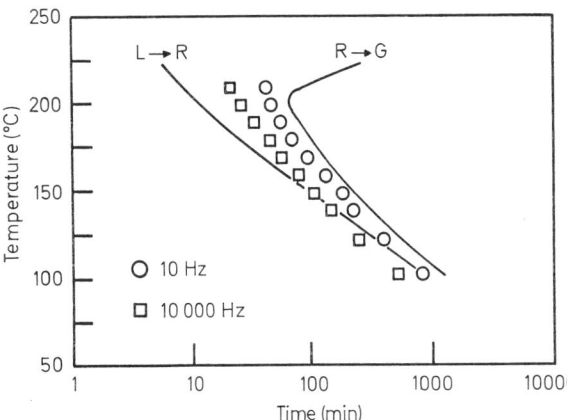

Fig. 26. Time-temperature-transformation diagram for the system EPON 825/DDS showing times to reach dipole loss peaks at 10 and 10,000 Hz. L-R denotes gelation; R-G denotes vitrification. (Reprinted from Ref. [47] with permission of the Society for the Advancement of Material and Process Engineering)

the relaxation data for rubber-modified epoxies, and found that β was sensitive to the rubber content of the resin.

The relation between the changing relaxation time distribution and the molecular structure of the curing system has not been determined. There is, however, one important experimental implication to these observations. Because both $\varepsilon_r$ and β change during cure, one must actually measure the complete frequency dependence of the $\varepsilon'$ and $\varepsilon''$ at a given state of cure in order to characterize the relaxation time distribution. Measurements made at one frequency throughout the cure process, when plotted on a Cole-Cole diagram, do not necessarily provide information about the relaxation time distribution.

## 4.5 Conductivity

This Section addresses the effects of temperature and cure on the ionic conductivity. The starting point is the assumption that the ions involved are primarily impurities, such as residual sodium and chloride ions, and that their concentrations do not change appreciably during cure. This appears to be a good assumption for epoxy resins [24]. It permits conductivity changes to be interpreted in terms of changes in ion mobility, which can, in turn, be related to the mobility of the polymer chains. However, these assumptions can be expected to fail when water is a product of the curing reaction, such as in phenolic resins and some polyimides (in which a protonic conduction mechanism has been suggested [81]). If the mobile ion content is changing, analysis of the conductivity requires knowing the concentration and mobility of each mobile species, introducing complexity analogous to that encountered in Section 4.3 in interpreting the relaxed permittivity.

As noted in Section 3.1, resistivity and viscosity are correlated prior to gelation, a result that has been evident since the earliest work in this field [1-4, 42, 43, 82-85]. However, as gelation is approached, the viscosity diverges, while the resistivity remains finite and varies continuously as cure proceeds. Figure 27 illustrates this behavior with data from Tajima and Crozier [83]. In Section 4.4, it was shown that the dipolar mobility tracks the polymer mobility through a WLF dependence on T and $T_g$. Similarly, one might expect that the conductivity (the reciprocal of resistivity) also tracks polymers mobility through a WLF Equation. The correlation between resistivity and viscosity prior to gelation, where the viscosity obeys a WLF Equation, supports this idea. Further direct evidence for the WLF behavior of the conductivity is reviewed below.

Sheppard [46] has studied the temperature dependence of the conductivity in a homologous series of DBEGA resins with n in the range from 0 to 12. The data are shown plotted in Arrhenius fashion in Fig. 28, with the solid lines representing the fit of the WLF Equation to the data. Similar temperature dependences of the conductivity have been reported in fully-cured epoxy systems [24, 86]. In both cases, and just as with the dipolar mobility, the driving force behind the observed temperature dependence is the mobility of the polymer as determined by the difference $T - T_g$. There are, however, differences in detail between the temperature dependences of the dipolar mobility and the conductivity. The first is that the $C_1$ constant Sheppard obtained for the conductivity is completely independent of the molecular

Dielectric Analysis of Thermoset Cure

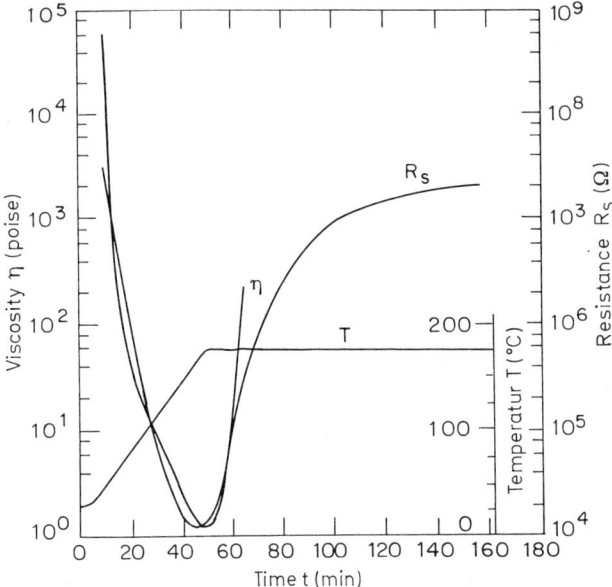

**Fig. 27.** Tajima's plot of viscosity and resistance during the cure of an epoxy resin. (Reprinted from Ref. [76] with permission of the Society of Plastics Engineers)

weight of the resin, which suggests that the free volume required for ion transport does not depend on the structural details of the molecular matrix. A further result is that unlike the dipolar mobility, for which the $C_2$ constant was nearly independent of n, the $C_2$ constant appropriate to the conductivity tracked $T_g$; that is, $T_g - C_2$ is nearly constant.

These results can be used as a guide to interpreting conductivity data during cure using the WLF Equation. As with the dipolar mobility, one would expect *the temperature dependence of the conductivity at fixed conversion to obey the WLF*

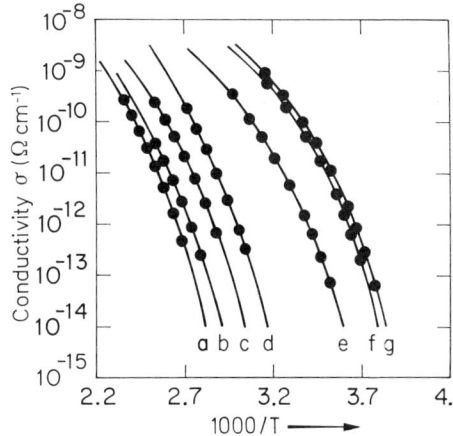

**Fig. 28.** Arrhenius plot of the conductivity for DGEBA epoxy resins of varying molecular weights. a) n = 12.1; b) 5.1; c) 3.4; d) 2.1; e) 0.6; f) n = 0.2; g) 0. (Reprinted from Ref. [46] with permission of the authors)

*Equation and the reference temperature to track the glass transition temperature as it increases during cure.* Correlated measurements of conductivity, chemical conversion, and $T_g$ have not been reported to verify this prediction, but experiments along these lines are under way at several laboratories.

Even in the absence of direct confirming evidence, we can use the WLF approach to explain a number of results already in the literature. Figure 29 illustrates schematically how the conductivity would vary with $1/T$ at two different extents of chemical conversion. The arrow indicates the pathway during an isothermal cure, in which the material starts at higher conductivity, where the slope against $1/T$ is relatively low, and moves toward lower conductivity, where the slope is larger. The slope represents the effective "activation energy" for conductivity which would be extracted from an Arrhenius plot of conductivity over a limited temperature range. The graph shows that this apparent activation energy would increase during cure, a result first reported by Warfield and Petree in 1959 [57], and subsequently analyzed by Sheppard [35]. Conceptually, as $T_g$ increases during cure toward the cure temperature, one is moving down the WLF curve to lower conductivity and steeper slope, hence, higher apparent activation energy.

This same idea helps explain the "knee" in the conductivity-versus-time data noted by many workers, but specifically interpreted as gelation by Acitelli, et al. [60]. The data are shown in Fig. 30a, including Acitelli's observation of the coincidence between the time to reach gelation and the time at which the extrapolated linear regions intersect. We now understand this coincidence; it is not related to gelation, but results from the autocatalyzed reaction kinetics coupled with the specific properties of the WLF Equation, as explained below.

When time rather than temperature is the variable, the reaction kinetics must be considered. As noted earlier, the epoxy-amine cure is autocatalyzed, and consequently, the maximum rate of reaction occurs at about 30 to 40% conversion. Therefore, the time rate of change of $T_g$ also reaches a maximum at these conversions, as demonstrated by Huguenin and Klein [48]. Furthermore, as cure proceeds, the difference $T - T_g$ decreases, and the conductivity becomes more sensitive to changes in $T_g$, as illustrated in Fig. 29.

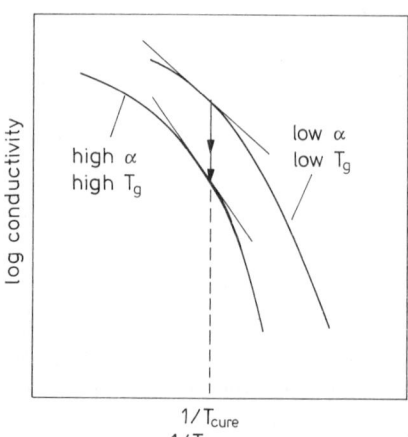

**Fig. 29.** Schematic illustration of the temperature dependence of the conductivity during two points of an isothermal thermoset cure. (Reprinted from Ref. [35] with permission of the Massachusetts Institute of Technology)

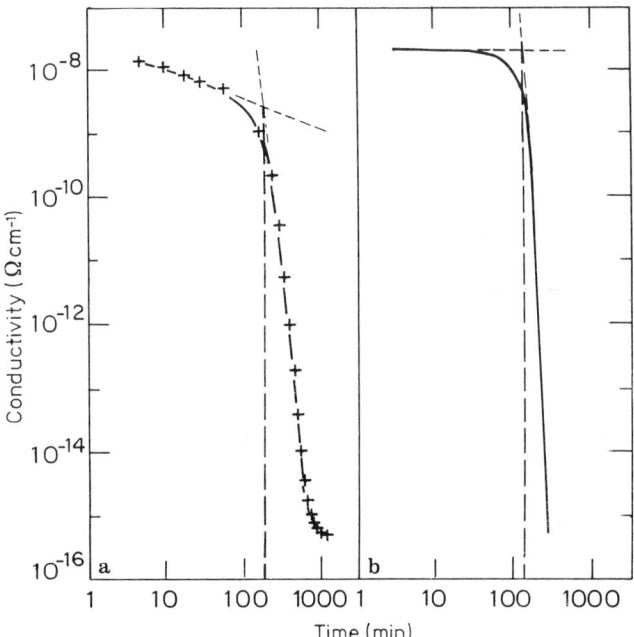

**Fig. 30a.** Conductivity data of Acitelli for the 57 °C cure of a low molecular weight DGEBA resin with m-phenylene diamine. (Reprinted from Ref. [60] with the permission of the publisher Butterworth Scientific, Ltd.). **b)** Simulation of the data of Figure 30a using a chemical kinetic model to determine $T_g$ versus time and a WLF Equation to determine the conductivity. (Reprinted from Ref. [35] with permission of the Massachusetts Institute of Technology)

The combined effect of these two mechanisms will lead to three regions of conductivity behavior in an isothermal cure. Early in cure, where the reaction rate is low and the sensitivity to $T_g$ changes is small, the conductivity decreases relatively slowly with cure time. Later in cure, both the reaction rate and rate of change of $T_g$ with time reach maximum values, while the sensitivity to $T_g$ changes increases, leading to a rapidly changing conductivity with time. Finally, towards the end of cure, the reaction becomes diffusion controlled, so the change in $T_g$ with time is very small, and the the conductivity versus time curve levels out. These three regions are clearly evident in the data of Fig. 30a. We have simulated the first two of these regions [35] by first estimating the variation of $T_g$ with time and then obtaining the conductivity from the WLF Equation using Sheppard's $C_1$ and $C_2$ constants for DGEBA resins [46]. The $T_g$ variation was determined by combining the autocatalyzed kinetics [61,63] with an empirical relation between $T_g$ and $\alpha$ proposed by Di-Bennedetto, and applied to epoxy systems by Adabbo and Williams [87] and by Enns and Gillham [78]. The resulting conductivity versus time is plotted in Fig. 30b, which shows the knee of the Acitelli data at the appropriate time. Since the model used for the simulation makes no reference to gelation, we assert that the coincidence of the knee with the gel time is fortuitous. This model also shows why it is inappropriate to attempt to equate time rates of change of dielectric parameters, such as conductivity, directly to time rates of change of chemical conversion (see also Sect. 4.2).

Blocking-layer effects are also sensitive to the conductivity change during cure. As demonstrated in Section 3.2, electrode polarization can lead to a Cole-Cole diagram that is similar to the Debye Equation for dipole orientation, but with much larger apparent permittivities. If the thickness of the blocking layer does not vary during cure (which will be the case if the ion content remains fixed), the "relaxation time" associated with this polarization is proportional to the resistivity of the medium, and, hence, should track viscosity early in cure. The present authors reported such a correlation [54]; Zukas et al., [79] have subsequently confirmed that the time to the loss peak associated with electrode polarization tracks the "isoviscous" event in torsional braid analysis.

Finally, from a pragmatic point of view, the conductivity is a very sensitive probe of cure. Not only does it become increasingly sensitive to changes in $T_g$ as the end of cure is reached, but it can almost always be measured, even in the presence of dipolar effects, by decreasing the measurement frequency. The conductivity is equally sensitive to small decreases in $T_g$ that result from degradation, a result first noted by Warfield [58]. Applications of conductivity measurement to cure studies are reviewed in Section 5.

# 5 Applications

In Section 4, we have examined, from a fundamental point of view, how temperature and cure affect the dielectric properties of thermosetting resins. The principal conclusions of that study were (1) that conductivity (or its reciprocal, resistivity) is perhaps the most useful overall probe of cure state, (2) that dipolar relaxations are associated with the glass transition (i.e., with vitrification), (3) that correlations between viscosity and both resistivity and dipole relaxation time are expected early in cure, but will disappear as gelation is approached, and (4) that the relaxed permittivity follows chemical changes during cure but is cumbersome to use quantitatively.

This Section presents a cross-referenced bibliography on the application of dielectric property measurements to thermosetting materials. Our literature search identified almost 200 papers with some relevance to the subject. Of these, we have selected about 70 for inclusion in this section. These papers provide particularly useful application examples, or provide data typical of the particular material or application which future investigators can use for comparisons with their own results.

The bibliography is presented in the form of two Tables. Table 1 references the papers by the material system under study; Table 2 references the papers by the application involved and/or by the type of data presented. As is evident in Table 1, most of the literature involves epoxies, and most of the applications to composites are either for epoxy-glass or epoxy-graphite systems.

In the Applications categories of Table 2, except for those papers identified as presenting post-cure results, all of the papers involve curing. No differentiation was made as to isothermal or ramped cure, since both types of data would be of importance to any particular resin system. The heading "Cure Rate and/or Catalyst Studies" includes those papers in which explicit correlations between cure temperature or catalyst concentration are presented, whereas the heading "General

Process Monitoring" includes papers where, for the most part, commercial resin or prepreg systems are carried through nominal cure cycles.

The Data Type categories of Table 2 are used to identify those papers either where quantitative results on the indicated dielectric property have been obtained, or where the data is typical of the system of interest.

**Table 1.** Bibliographic references by materials studied

## RESINS

Pure Epoxies [46, 69]
DGEBA, Amine cured [41, 47, 54, 55, 58, 60, 67, 68, 70, 71, 73, 93, 94]
DGEBA, Anhydride cured [49, 50, 51, 52, 55, 56, 58, 96, 104]
DGEBA, Other curing agents [41, 44, 74, 75, 84, 85, 92, 105, 112]
Tetrafunctional Aerospace resins [39, 40, 41, 42, 43, 44, 45, 74, 75, 79, 82, 83, 84, 91, 95, 95, 97, 98, 99, 100, 101, 118]
Other epoxies including rubber-modified and novolac resins [37, 40, 80, 86, 88, 89, 90, 91, 96, 97, 101, 102, 103, 105]
Polyesters [57, 64, 65, 106, 107, 108, 110]
Phenolics [2, 5, 6, 109]
Polyimides [41, 45, 84, 95, 101, 111, 112]
Polyurethanes [69, 113]
Rubber [115, 116]
Shellac and other resins [114, 117]

## COMPOSITES

Epoxy-Glass [40, 43, 44, 88, 89, 90, 91, 96, 98]
Epoxy-Graphite [40, 42, 44, 45, 82, 83, 91, 95, 98, 99, 100, 118]
Epoxy with other fibers and fillers [102, 103, 105]
Polyimide-Graphite [112]

**Table 2.** Bibliographic references by Application and data type

## APPLICATION

Cure rate and/or catalyst studies [37, 39, 40, 41, 56, 57, 58, 60, 64, 65, 67, 68, 69, 70, 71, 79, 91, 93, 94, 95, 97, 100, 105, 108, 111, 112]
Correlations with mechanical properties [39, 42, 43, 47, 54, 55, 60, 67, 68, 70, 74, 75, 79, 82, 83, 85, 88, 95, 100, 107, 115]
Moisture effects [90, 92, 100, 109, 118]
Aging effects [40, 83, 91, 95, 100, 102, 111, 112, 118]
General Process Monitoring [40, 41, 42, 43, 44, 45, 82, 83, 84, 86, 88, 89, 90, 91, 95, 96, 97, 98, 99, 100, 101] (bond lines), [110] (wire coating), [113, 116, 117]
Post-cure studies [49, 50, 51, 52, 56, 57, 58, 80, 84, 86, 92, 93, 94, 100, 103, 104, 109, 111, 113, 114]

## DATA TYPE

Conductivity and/or loss factor [2, 5, 6, 37, 39, 44, 46, 47, 49, 50, 54, 57, 58, 60, 64, 65, 67, 68, 69, 70, 74, 75, 79, 80, 82, 83, 84, 85, 86, 88, 89, 92, 93, 94, 97, 99, 103, 106, 107, 109, 112, 113, 114, 115, 116]
Permittivity [37, 39, 46, 47, 49, 50, 51, 52, 54, 55, 56, 67, 68, 69, 70, 71, 73, 74, 75, 79, 80, 84, 85, 92, 93, 94, 97, 103, 104, 108, 109, 112, 113, 114]
Dissipation Factor [40, 41, 42, 43, 45, 51, 52, 55, 56, 82, 83, 84, 89, 90, 91, 92, 95, 96, 98, 100, 102, 104, 105, 108, 110, 111, 112, 115, 117, 118]

*Acknowledgement*: The authors wish to acknowledge the research support of the United States Office of Naval Research during the preparation of this review, and the literature search assistance of Rebecca Miller, who received partial support through the MIT Undergraduate Research Opportunities Program.

# 6 Nomenclature

| | |
|---|---|
| $\alpha$ | extent of conversion |
| $\beta$ | distribution parameter |
| $\tan \delta$ | bulk material loss tangent |
| $\tan \delta_x$ | experimental loss tangent (equivalent to D) |
| $\varepsilon_0$ | permittivity of free space, $8.85 \times 10^{-14}$ F/cm |
| $\varepsilon_r$ | relaxed permittivity (relative to $\varepsilon_0$) |
| $\varepsilon_u$ | unrelaxed permittivity (relative to $\varepsilon_0$, equivalent to $\varepsilon_\infty$) |
| $\varepsilon'$ | bulk permittivity (relative to $\varepsilon_0$) |
| $\varepsilon'_x$ | experimental permittivity (relative to $\varepsilon_0$) |
| $\varepsilon'_d$ | dipole contribution to the relative permittivity |
| $\varepsilon_\infty$ | infinite-frequency relative permittivity (equivalent to $\varepsilon_u$) |
| $\varepsilon''$ | bulk loss factor (relative to $\varepsilon_0$) |
| $\varepsilon''_x$ | experimental loss factor (relative to $\varepsilon_0$) |
| $\eta$ | viscosity |
| $\mu$ | dipole moment |
| $\varrho$ | resistivity (equivalent to $1/\sigma$) |
| $\sigma$ | conductivity (equivalent to $1/\varrho$) |
| $\tau_d$ | dipole relaxation time |
| $\varphi$ | phase difference between voltage and current |
| $\omega$ | angular frequency |
| $\Omega$ | ohms |
| A | area of parallel plate electrodes |
| C | equivalent circuit representation of bulk sample capacitance |
| $C_b$ | blocking-layer capacitance |
| $C_L$ | capacitance of microdielectrometer floating gate |
| $C_x(\omega)$ | experimentally measured sample capacitance |
| $C_1$ | constant in Williams-Landel-Ferry (WLF) equation |
| $C_2$ | constant in Williams-Landel-Ferry (WLF) equation |
| D | dissipation factor (equivalent to $\tan \delta_x$) |
| $H(\omega)$ | microdielectrometer transfer function |
| i(t) | time-varying current |
| I | complex amplitude of i(t) |
| L | spacing of parallel plate electrodes |
| $N_i$ | concentration of $i^{th}$ ion |
| $q_i$ | charge of $i^{th}$ ion |
| Q(t) | time-varying charge |
| $r_i$ | radius of $i^{th}$ ion |
| R | equivalent circuit representation of bulk sample resistance |

$R_x(\omega)$   experimentally measured sample resistance
$t_b$   blocking-layer thickness
$T_g$   glass transition temperature
$u_i$   mobility of $i^{th}$ ion
$v(t)$   time-varying voltage
$V$   complex amplitude of v(t)
$Y(\omega)$   experimentally measured admittance

# 7 References

1. Kienle, R. H., Race, H. H.: The electrical, chemical and physical properties of alkyd resins, Trans. Electrochem. Soc., 65, 87 (1934)
2. Manegold, E., Petzoldt, W.: Die elektrische Leitfähigkeit während der Reaktion von sauren bzw. alkalischen phenol-formaldehyd-Gemischen, Kolloid Z., 95, 59 (1941)
3. Lomakin, B. A., Gussewa, N. J.: Kunststoffe, 29, 145 (1939)
4. Lomakin, B. A., Gussewa, N. J.: Anwendung von Leitfähigkeits-, Viskositäts- und Refraktionsmessungen zum Studium der Harzbildung, Plast. Massi 2, 281 (1937)
5. Fineman, M. N., Puddington, I. E.: Kinetics of cure of resol resins, Ind. Eng. Chemistry, 39, 1288 (1947)
6. Fineman, M. N., Puddington, I. E.: Measurements of cure of some thermosetting resins, Can. J. Res., Sec. B., 25, 101 (1947)
7. Coln, M. C. W., Senturia, S. D.: The application of linear system theory to parametric microsensors, p. 118, Proc. Transducers '85, 1985
8. Mopsik, F. I.: Precision time-domain dielectric spectrometer, Rev. Sci. Instr. 55, 79 (1984)
9. Aukward, J. A., Warfield, R. W.: Monitoring device for investigation of encapsulating resins, Rev. Sci. Inst., 27, 413 (1956)
10. Armstrong, R. J.: Method of laminating employing measuring the electrical impedance of a thermosetting resin, U.S. Patent No. 3,600,247
11. Mouayad, L.: Monitoring of transformer oil using microdielectric sensors, M.S. Thesis, Massachusetts Institute of Technology, 1985, unpublished
12. Lee, H. L.: Optimization of a resin cure sensor, M.S. Thesis, Massachusetts Institute of Technology, 1982, unpublished
13. Senturia, S. D., Garverick, S. L.: Method and apparatus for microdielectrometry, U.S. Patent No. 4,423,371
14. Sheppard, Jr., N. F., Garverick, S. L., Day, D. R., Senturia, S. D.: Microdielectrometry: a new method for in-situ cure monitoring, p. 65, Proc. 26th SAMPE Symposium, 1981
15. Senturia, S. D., Sheppard, Jr., N. F., Lee, H. L., Marshall, S. B.: Cure monitoring and control with combined dielectric/temperature probes, SAMPE J., 19, 22 (1983)
16. Aukward, J. A., Warfield, R. W., Petree, M. C.: Change in electrical resistivity of some high polymers during isothermal polymerization, J. Polym. Sci., 27, 199 (1958)
17. Anonymous: Dielectrics improve control of laminate cure time, Plastics Technology, 15, 19 (1969)
18. Hewlett-Packard Co., Palo Alto, CA: Manual for HP4192A Low-Frequency Impedance Analyzer
19. GenRad Co., Concord MA: Manual for 1689 Digibridge
20. Micromet Instruments, Inc., Cambridge, MA: The Eumetric System II Microdielectrometer, Product Bulletin, 1985
21. McCrum, N. G., Read, B. E., Williams, G.: Anelastic and dielectric effects in polymeric solids, New York, Wiley and Sons 1967
22. Hedvig, P.: Dielectric spectroscopy of polymers, New York, Wiley and Sons 1977
23. Smyth, C. P.: Dielectric behavior and structure, New York, McGraw-Hill 1955
24. Fava, R. A., Horsfield, A. E.: The interpretation of electrical resistivity measurements during epoxy resins cure, Brit. J. Appl. Phys. (J. Phys. D), Ser. 2, 1, 117 (1968)

25. Lee, H., Neville, K.: Handbook of epoxy resins, New York, McGraw-Hill 1967
26. Wright, W. W.: Characterisation of a Bisphenol-A epoxy resin, Brit. Polym. J., *15*, 224 (1983)
27. Bockris, J. O'M., Reddy, A. K. N.: Modern electrochemistry, Volume 1, Section 4.4, New York, Plenum Press 1970
28. Debye, P.: Polar molecules, New York, Chemical Catalog Co. 1929
29. Cole, K. S., Cole, R. H.: Dispersion and absorption in dielectrics I. Alternating current characteristics, J. Chem. Phys., *9*, 341 (1941)
30. Davidson, D. W., Cole, R. H.: Dielectric relaxation in glycerine, J. Chem. Phys., *18*, 1417 (1950)
31. Williams, G., Watts, D. C.: Non-symmetrical dielectric relaxation behavior arising from a simple empirical decay function, Trans. Farad. Soc., *66*, 80 (1970)
32. Williams, G., Watts, D. C., Dev, S. B., North, A. M.: Further considerations of non-symmetrical dielectric relaxation behavior arising from a simple empirical decay function, Trans. Farad. Soc., *67*, 1323 (1971)
33. Lindsey, C. P., Patterson, G. D.: Detailed comparison of the Williams-Watts and Cole-Davidson functions, J. Chem. Phys. *73*, 3348 (1980)
34. Moynihan, C. T., Boesch, L. P., Laberge, N. L.: Decay function for the electric field relaxation in vitreous ionic conductors, Phys. Chem. Glasses, *14*, 122 (1973)
35. Sheppard, Jr., N. F.: Dielectric analysis of the cure of thermosetting epoxy/amine systems, Ph.D. Thesis, Massachusetts Institute of Technology, 1985, unpublished
36. Johnson, J. F., Cole, R. H.: Dielectric polarization of liquid and solid formic acid, J. Am. Chem. Soc., *73*, 4536 (1951)
37. Adamec, V.: Electrical properties of an epoxy resin during and after curing. J. Polym. Sci., *10*, 1277 (1972)
38. Day, D. R., Lewis, T. J., Lee, H. L., Senturia, S. D.: The role of boundary layer capacitance at blocking electrodes in the interpretation of dielectric cure data in adhesives, J. Adhesion, *18*, 73 (1985)
39. Zukas, W. X., MacKnight, W. J., Schneider, N. S.: Dynamical mechanical and dielectric properties of an epoxy resin during cure, in C. A. May (ed.), Chemorheology of Thermosetting Polymers, ACS Symposium Series, *227*, 223 (1983)
40. May, C. A.: Dielectric measurements for composite cure control — two case studies, p. 108, Proc. 20th SAMPE Symp., 1975
41. May, C. A.: Composite cure studies by dielectric and calorimetric analyses, p. 803, Proc. 21st SAMPE Symp., 1976
42. Lawless, G. W.: High temperature dielectric studies of epoxy resin, Polym. Eng. Sci., *20*, 546 (1980)
43. Baumgartner, W. E., Ricker, T.: Computer assisted dielectric cure monitoring in material quality and cure process control, SAMPE J., *19*, 6 (1983)
44. Bidstrup, W. B., Sheppard, Jr., N. F., Senturia, S. D.: Monitoring of laminate cure with microdielectrometry, SPE Tech. Papers, *31*, 324 (1985)
45. Pike, R. A., Douglas, F. C., Wisner, G. R.: Electrical dissipation factor; guide to molding quality graphite/resin composites, Polym. Eng. Sci., *11*, 502 (1971)
46. Sheppard, Jr., N. F., Senturia, S. D.: WLF dependence of the dielectric properties of DGEBA epoxy resins, J. Polym. Sci., Polym. Phys. Ed., submitted for publication
47. Sheppard, Jr., N. F., Coln, M. C. W., Senturia, S. D.: A dielectric study of the time-temperature-transformation (TTT) diagram of DGEBA epoxy resins cured with DDS, p. 1243, Proc. 29th SAMPE Symp., 1984
48. Huguenin, F. G. A. E., Klein, M. T.: Intrinsic and transport-limited epoxy-amine cure kinetics, Ind. Eng. Chem. Prod. Res. Dev., *24*, 166 (1985)
49. Shito, N., Sato, M.: Electrical and mechanical properties of anhydride-cured epoxy resins, J. Polym. Sci. Part C, *16*, 1069 (1967)
50. Shito, M.: Specific volume, dielectric properties, and mechanical properties of cured epoxy resins in the glass-transition region, J. Polym. Sci. Part C, *23*, 569 (1968)
51. Blyakhman, Ye. M., Borisova, T. I., Levitskaya, Ts. M.: Dielectric relaxation in epoxide resins hardened with different anhydrides, Polym. Sci. USSR, *12*, 1756 (1970)
52. Blyakhman, Ye. M., Borisova, T. I., Levitskaya, Ts. M.: Molecular mobility study on set epoxy resins of different molecular weights, using the dielectric method, Polym. Sci. USSR, *12*, 2602 (1970)

53. Williams, M. L., Landel, R. F., Ferry, J. D.: The temperature dependence of relaxation mechanisms in amorphous polymers and other glass-forming liquids, J. Am. Chem. Soc., 77, 3701 (1955)
54. Sheppard, Jr., N. F., Day, D. R., Lee, H. L., Senturia, S. D.: Microdielectrometry, Sensors & Actuators, 2, 263 (1982)
55. Delmonte, J.: Electrical properties of epoxy resins during polymerization, J. Appl. Polym. Sci., 2, 108 (1959)
56. Olyphant, Jr., M.: Effect of cure and aging on dielectric properties, Suppl. p. 12, Proc. 6th IEEE Electrical Insulation Conf., 1965
57. Warfield, R. W., Petree, M. C.: The use of electrical resistivity in the study of the polymerization of thermosetting polymers, J. Polym. Sci., 37, 305 (1959)
58. Warfield, R. W., Petree, M. C.: A study of the polymerization of epoxide polymers by electrical resistivity techniques, Polymer, 1, 178 (1960)
59. Kagan, G. T., Moshinskii, L. Ya., Nesolyonaya, L. G., Marina, D. I., Romantsevich, M. K.: Study of kinetics of setting of epoxide resins with anhydrides, Vysokomol. Soedin., $A10$, 62 (1968)
60. Acitelli, M. A., Prime, R. B., Sacher, E.: Kinetic of epoxy cure: (1) the system Bisphenol-A diglycidyl ether/m-phenylenediamine, Polymer, 12, 335 (1971)
61. Prime, R. B.: Thermosets, in Edith Turi (ed.), Thermal Characterization of Polymeric Materials, New York, Academic Press 1981
62. Schechter, L., Wynstra, J., Kurkjy, R. P.: Glycidyl ether reactions with amines, Ind. Eng. Chem., 48, 94 (1956)
63. Sourour, S., Kamal, M. R.: Differential scanning calorimetry of epoxy cure: isothermal cure kinetics, Thermochimica Acta, 14, 41 (1976)
64. Judd, N. C. W.: Investigation of the polymerization of an unsaturated polyester system by a resistivity technique, J. Appl. Polym. Sci., 9, 1743 (1965)
65. Learmonth, G. S., Pritchard, G.: Electrical resistivity and crosslinking in thermosetting resins, Ind. Eng. Chem. Prod. Prod. Res. Dev., 8, 124 (1969)
66. Fisch, W., Hofmann, W.: Chemischer Aufbau von gehärteten Epoxydharzen, Makromol. Chem., 44, 8 (1961)
67. Huraux, C., Sellaimia, A.: Variation des proprietes dielectriques accompagnant le durcissement de resines epoxydes, C.R. Acad. Sc. Paris, 277, Serie B, 497 (1973)
68. Huraux, C., Sellaimia, A.: Sur la variation des proprieties dielectriques et la cinetique du durcissement de resines epoxides, C.R. Acad. Sc. Paris 277, Serie B, 691 (1973)
69. Dandurant, D., Huraux, C.: Etude de mecanisme de durcissement de resines au moyen de spectres de relaxation dielectrique, Prace Naukowe Instytutu Podstaw Elektrotechniki i Electrotechnologii Politechniki Wrocławskiej, 16, 275 (1977)
70. Soualmia, A., Huraux, C., Despax, B.: Relation entre les parametres dielectriques et la cinetique de reaction de polymerisation en phase liquide, Makromol. Chem., 183, 1803 (1982)
71. Sheppard, Jr., N. F., Senturia, S. D.: Chemical interpretation of the relaxed permittivity during epoxy resin cure, SPE Tech. Papers, 31, 321 (1985)
72. Williams, G.: Dielectric information on chain mobility, chain configuration, and an orderdisorder transition in amorphous poly(acetaldehyde), Trans. Farad. Soc., 59, 1397 (1962)
73. Sheppard, J., N. F., Senturia, S. D.: Molecular contributions to the dielectric permittivity of unreacted epoxy/amine mixtures, p. 22a, Technical Program Summary, Adhesion Society Annual Meeting, Savannah, GA, 1985
74. Lane, J. W., Bachmann, M. S., Seferis, J. C.: Monitoring of matrix property changes during composite processing, SPE Tech. Papers, 31, 318 (1985)
75. Lane, J. W., Bachmann, M. S., Seferis, J. C.: Dielectric Studies of the cure of epoxy matrix systems, J. Appl. Polym. Sci., in press
76. Tajima, Y. A., Crozier, D.: Thermokinetic modeling of an epoxy resin I. Chemoviscosity, Polym. Eng. Sci., 23, 186 (1983)
77. Apicella, A., Nicolais, L., Nobile, M. R., Castiglione-Morelli, M. A.: Effect of processing variables on the durability of epoxy resins for composite systems, Comp. Sci. Tech., in press

78. Enns, J. B., Gillham, J. K.: Time-temperature-transformation (TTT) cure diagram: modeling the cure behavior of thermosets, J. Appl. Polym. Sci., 28, 2567 (1983)
79. Zukas, W. X., Schneider, N. S., MacKnight, W. J.: Dielectric and dynamic mechanical monitoring of epoxy resin cure, Polym. Prepr., 25 (2), 205 (1984)
80. Daly, J., Pethrick, R. A.: Rubber-modified epoxy resins: 2. Dielctric and ultrasonic relaxation studies, Polymer, 22, 37 (1981)
81. Sacher, E.: The DC conductivity of polyimide film, p. 33, Proc. IEEE Conf. on Elec. Ins. Dielec. Phen., 1976
82. Carpenter, J. F.: Instrumental techniques for developing epoxy cure cycles, p. 783, Proc. 21st SAMPE Symposium, 1976
83. Tajima, Y. A.: Monitoring cure viscosity of epoxy composite, Polym. Comp., 3, 162 (1982)
84. Kranbuehl, D. E.: Dynamic dielectric characterization of thermosets and thermoplastics using intrinsic cariables, p. 1251, Proc. 29th SAMPE Symp., 1984
85. Kranbuehl, D. E., Delos, S., Yi, E., Mayer, J., Hou, T., Winfree, W.: Correlation of dynamic dielectric measurements with viscosity in polymeric resin systems, p. 638, Proc. 30th SAMPE Symp., 1985
86. McGowan, E. J., Mathes, K. N.: Measurement of cure in casting resins by means of electrical properties, p. 17, Proc. IEEE Conf. on Elec. Insul., 1962
87. Adabbo, H. E., Williams, R. J. J.: The evolution of thermosetting polymers in a conversion-temperature phase diagram, J. Appl. Polym. Sci. 27, 1327 (1982)
88. Armstrong, R. J.: Laminating control system, Solid State Technology, p. 50, Nov. 1969
89. Kim, D. H.: Application of the dielectric analysis to polymeric materials control, Proc. TTCP-3 Critical Review; Techniques for the Characterization of Polymeric Materials, AMMRC MS 77-2, AD-36082, 1977
90. Dixon, R. R.: Measuring moisture by dielectric analysis, SPE Tech. Papers, 25, 835 (1979)
91. May, C. A., Hadad, D. K., Browning, C. E.: Physiochemical quality assurance methods for composite matrix resins, Polym. Eng. Sci., 19, 545 (1979)
92. Standish, J. V., Leidheiser, Jr., H.: The effect of water on the dielectric properties of a corrosion-protective epoxy-polyamide coating, ACS Div. Org. Coat. Plast. Chem. Prepr. 43, 565 (1980)
93. Day, D. R.: Effects of stoichiometric mixing ratio on epoxy cure — dielectric analysis, SPE Tech. Papers, 31, 327 (1985)
94. Senturia, S. D., Sheppard, Jr., N. F., Lee, H. L., Day, D. R.: In-situ measurement of the properties of curing systems with microdielectrometry, J. Adhesion, 15, 69 (1982)
95. May, C. A., Dusi, M. R., Fritzen, J. S., Hadad, D. K., Maximovich, M. G., Thrasher, K. G., Wereta, Jr., A.: A rheological and chemical overview of thermoset curing, in C. A. May (ed.), Chemorheology of Thermosetting Polymers, ACS Symposium Series 227, 1 (1983)
96. Chottiner, J., Sanjana, Z. N., Kodani, M. R., Lengel, K. W., Rosenblatt, G. B.: Monitoring cure of large autoclave molded parts by dielectric analysis, p. 77, Proc. 26th SAMPE Symp., 1981
97. Sanjana, Z. N., Selby, R. L.: Monitoring cure of epoxy resins using a microdielectrometer, p. 1233, Proc. 29th SAMPE Symp., 1984
98. Allen, J. D.: In-process dielectric monitoring of polymeric resin cure, p. 270, Proc. 20th SAMPE Symp., 1975
99. Crabtree, D. J.: Ion graphing as an in-process cure monitoring procedure for composite and adhesively bonded structures, p. 636, Proc. 22nd SAMPE Symp., 1977
100. Sanjana, Z. N., Schaefer, W. H., Ray, J. R.: Effect of aging and moisture on the reactivity of a graphite epoxy prepreg, Polym. Eng. Sci., 21, 474 (1981)
101. Wereta, Jr., A., May, C. A.: Dielectric monitoring of the bonding process, J. Adhesion, 12, 317 (1981)
102. Pike, R. A., Lamm, F. P., Pinto, J. P.: Factors affecting the processing of epoxy film adhesives: 1. Room temperature aging, J. Adhesion, 12, 143 (1981)
103. La Mantia, F. P., Schifani, R., Acierno, D.: Effect of a filler on the dielectrid properties of an an epoxy resin, J. Appl. Polym. Sci., 28, 3075 (1983)
104. Delmonte, J.: Electrical properties of modified epoxy resins, Modern Plastics, 35, 152 (1958)

105. Sanjana, Z. N., Selby, R. L.: The use of dielectric analysis to study the cure of a filled epoxy resin, IEEE Trans. Elec. Ins., *16*, 496 (1981)
106. Pais, J. C., Griffiths, V. S.: The effect of chemical bonding on current flow in thermosetting polymers, p. 5, Proc. IEEE Conf. on Elec. Insul. and Dielec. Phen., 1968
107. Learmonth, G. S., Tomlinson, F. M., Czerski, J.: Cure of polyester resins, I., J. Appl. Polym. Sci., *12*, 403 (1968)
108. Learmonth, G. S., Pritchard, G.: Dielectric relaxation and crosslinking in unsaturated polyester resins, J. Appl. Polym. Sci., *13*, 1219 (1969)
109. Van Beek, L. K. H.: Dielectric behavior of curing phenolic-formaldehyde resins, J. Appl. Polym. Sci., *8*, 2843 (1964)
110. Sanjana, Z. N.: A study of wire enamel cure using dielectric analysis, Proc. IEEE Conf. Electr. and Electronic Insul., *15*, 150 (1981)
111. Kranbuehl, D. E., Delos, S., Jue, P. K.: Dynamic dielectric characterization of the cure process: LARC-160 SAMPE J., *19*, 18 (1983)
112. Kranbuehl, D. E., Delos, S., Yi, E.: »ynamic dielctric analysis: nondestructive material evaluation and cure cycle monitoring, SPE Tech. Papers, *31*, 311 (1985)
113. Krishna, J. G., Josylu, O. S., Sobhanadri, J., Subrahmaniam, R.: Dielectric behavior of isocyanate-terminated polymers, J. Phys. D: Appl. Phys. *15*, 2315 (1982)
114. Goswami, D. N.: The dielectric behavior of natural resin shellac, J. Appl. Polym. Sci., *23*, 529 (1979)
115. Khastgir, D., Ghoshal, P. K., Das, C. K.: Dielectric behavior and sulphur vulcanization of the crosslinked butyl (XL-50) rubber, Polymer, *24*, 617 (1983)
116. Hinkley, J. A.: Monitoring of a rubber cure by microdielectrometry, U. S. Naval Research Lab Memorandum Report 5300, March 30, 1984
117. Wereta, Jr., A., May, C. A.: Dielectric cure investigation of acetylene terminated oligomers, ACS Div. Org. Coat. Plast. Chem. Prepr., *38*, 679 (1978)
118. Sanjana, Z. N.: Overage indicators for prepreg products, SAMPE J., *16*, 5 (1980)


Editor: K. Dušek
Received October 31, 1985

# Epoxy-Aromatic Amine Networks in the Glassy State Structure and Properties

Eduard F. Oleinik
Institute of Chemical Physics, Academy of Sciences, Moscow, USSR

*Properties of networks, prepared by curing of diglycidyl ethers of bisphenols with some aromatic amines, have been considered. Network polymers, obtained from mixtures with different reactant ratios and at different curing temperatures ($T_{cure}$), have been chemically characterized (curing conversion, concentration of chemical crosslinks) in detail at all stages of the cure process. Polymer properties, i. e. mechanical features such as rigidity and deformability and thermal features such as $T_g$, heat capacity, and coefficient of thermal expansion, have been analyzed and interpreted in terms of chemical and physical (packing density) structure. Plastic deformation and some fracture peculiarities of network glasses have been examined. In a few cases, the properties of these polymers in the rubbery state have been considered. A comparison of the structure and properties of network and linear polymeric glasses has been made. It has been shown that, in the formation of the structure and properties of these polymers, an important role is played by vitrification during cure. Polymers prepared at low $T_{cure}$ usually show higher mechanical properties which, however, are not due to the high density of their glassy state. The peculiarities of the mechanical behaviour and vitrification of polymers isothermally cured at different $T_{cure}$ have been considered.*

| | |
|---|---|
| **Abbreviations and Symbols** | 50 |
| **1 Introduction** | 51 |
| **2 Important Features of Chemistry and Topology of Epoxy Aromatic Amine Networks** | 52 |
| 2.1 Reactions in the Cure Process | 52 |
| 2.2 Vitrification During Cure | 55 |
| 2.3 Network Topology | 56 |
| 2.4 Main Results of Network Topological Analysis | 59 |
| **3 Structural Features of the Glassy State** | 60 |
| 3.1 Packing Density | 60 |
| 3.2 Heat Capacity $C_p$ and Thermal Expansion Coefficient | 61 |
| 3.3 Hydrogen Bonds in Epoxy Networks | 65 |
| 3.4 Local Molecular Motions | 66 |
| 3.4.1 Thermally Stimulated Discharge (TSD) Measurements | 66 |
| 3.4.2 Conformations of Aliphatic Chains | 69 |
| **4 Glass Transition Temperature** | 71 |
| 4.1 $T_g^\infty$ and the Chemical Structure of Networks | 72 |
| 4.2 Glass Transition Temperature and Chemical Side Reactions | 75 |
| **5 Mechanical Properties** | 75 |
| 5.1 Mechanical Properties in the Rubbery State | 75 |

      5.1.1 Experimental Results . . . . . . . . . . . . . . . . 75
      5.1.2 Conformational Elasticity of Diepoxide Chains . . . . . . . . . 76
      5.1.3 Calculation of the Elasticity Modulus . . . . . . . . . . . . . 77
  5.2 Mechanical Properties of the Glassy State . . . . . . . . . . . . . . 78
      5.2.1 Influence of the Chemical and Topological Structure on the
           Mechanical Behaviour of Networks . . . . . . . . . . . . . . 78
  5.3 Post-Cure Thermal Treatment and Mechanical Properties
      of the Networks. . . . . . . . . . . . . . . . . . . . . . . . . 82
  5.4 Plastic Deformation of Networks . . . . . . . . . . . . . . . . . 83

**6 Structure and Properties of Networks Cured at Low $T_{cure}$** . . . . . . . . . 86
  6.1 Quenching of the Cure Process . . . . . . . . . . . . . . . . . . 86
  6.2 Some Properties of Glassy Networks Prepared at Low $T_{cure}$ . . . . . . 88
      6.2.1 Glass Transition Temperature $T_g^{exp}$ . . . . . . . . . . . . . 88
      6.2.2 Mechanical Properties of Samples Prepared at Low $T_{cure}$ . . . . 92
  6.3 Structure of Network Glasses Prepared at Low $T_{cure}$ . . . . . . . . . 93

**7 Fracture of Epoxy-Aromatic Amine Networks** . . . . . . . . . . . . . . 94
  7.1 Fracture in the Rubbery State . . . . . . . . . . . . . . . . . . . 94
  7.2 Fracture in the Glassy State . . . . . . . . . . . . . . . . . . . . 95

**8 Conclusions** . . . . . . . . . . . . . . . . . . . . . . . . . . . . 96

**9 References** . . . . . . . . . . . . . . . . . . . . . . . . . . . . 97

## Abbreviations and Symbols

| | |
|---|---|
| $\alpha$ | reaction conversion |
| $\alpha^{dif}$ | diffusion conversion limit |
| $\alpha^{top}$ | topological conversion limit |
| AN | aniline |
| $\beta_g$ | cubic thermal expansion coefficient of the glassy state |
| $\beta_l$ | cubic thermal expansion coefficient of the liquid state |
| $\Delta\beta(T_g)$ | thermal expansion coefficient jump at $T_g$ |
| $C_p$ | heat capacity |
| $\Delta C_p(T_g)$ | heat capacity jump at $T_g$ |
| DGER | diglycidyl ether of resorcinol |
| DGEBA | diglycidyl ether of Bisphenol-A |
| DGEOPH | diglycidyl ether of orthophthalic acid |
| DGEPC | diglycidyl ether of pyrocathechol |
| DGEHHPH | diglycidyl ether of hexahydrophthalic acid |
| DADMOPHM | diaminodimethoxyphenylmethane |
| 4,4'-DADPhS | 4,4'-diaminodiphenyl sulfone |
| $E_{pol}$ | electric polarization field |
| E, E' | Young modulus; dynamic Young modulus |
| $\varepsilon$ | relative deformation |
| $\varepsilon_b$ | elongation at break |
| $\varepsilon^*$ | maximum local elongation |
| $\varepsilon_y$ | elongation at yield |

| | |
|---|---|
| G | shear modulus |
| $G_\infty$ | equilibrium shear modulus |
| $K_c$ | rigidity modulus of a network molecular chain |
| $K_{20}$ | packing density coefficient at 20 °C |
| $L_\perp, L_\parallel$ | length of plastic deformation zone in the front of crack tip ($\perp$ and $\parallel$ the tension direction) |
| $m$-PhDA | $m$-phenylenediamine |
| $\mu$ | dipole moment |
| N | number of dipoles |
| N(x) | number of crosslinks ($m$-PhDA molecules) with different |
| $x = 0 \div 4$ | connectivity (x) of the crosslink with entire network |
| P | initial molar ratio of reactive groups; $P = [NH]_0/[EP]_0$ |
| PhGE | phenylglycidyl ether |
| PETP | poly(ethylene terephthalate) |
| $P_0$ | saturation polarization |
| PS | polystyrene |
| PC | polycarbonate |
| Q | integral heat of cure reaction |
| $T_g^\infty$ | glass transition temperature of the network at the conversion $\alpha = \alpha^{top}$ |
| $T_g^{exp}$ | experimental glass transition temperature |
| $T_{cure}$ | cure temperature |
| $\Delta(T_g)$ | temperature interval of the glass transition |
| $\sigma_y$ | yield stress |
| $\tau_y$ | shear yield stress |
| $\tau_i$ | average waiting time of a breaking of chemical bond at the stress $\sigma_i$ |
| $\lambda = 1 + \varepsilon$ | |
| $U_0$ | activation energy |
| $V_{sp}$ | specific volume |

# 1 Introduction

Epoxy glasses have been widely utilized by different industries primarily as matrices for composites. Increasing interest in the most modern materials — advanced composites — where epoxies appear as the main type of matrices [1-5] led to the necessity of predicting structure — properties relationships for these polymers.

There are at least two reasons why these relationships have not yet been elucidated: the crosslinked nature of cured epoxies and their glassy state. The structure-properties relationship for the glassy state of matter, including polymers, is poorly understood and our knowledge of crosslinked glasses is even more insufficient.

In the literature, there exist many papers on the properties of epoxy glasses, but most of them are difficult to generalize due to incomplete characterization of the chemical and topological structure of the investigated samples. Sometimes, the difficulties of polymer characterization arise from the industrial origin of uncured resins and curing agents.

To receive basic results in the laboratory, we often need a good model system which can be quantitatively characterized at each step of the experimental study.

Ten years ago, good model systems were found for epoxy networks. These were polymers based on monomeric diglycidyl ethers of some bisphenols cured by simple aromatic amines (primarily m-phenylenediamine — mPhDA). Polymers based on these reactants satisfied the requirements for such a model system in several points:
— the chemistry of curing reaction and side reactions was carefully investigated [6,8];
— structural models of networks in the rubbery state were created [7,9,10]. The models gave a tool to quantitatively connect some features of network structure with their mechanical properties;
— all polymers showed a high level of mechanical properties in the glassy state [6,11,12–15].

This paper will deal mainly with this type of polymers, focussing on the glassy state of epoxy-aromatic amine-cured polymers. However, many aspects of network glasses cannot be understood without considering the chemistry of their formation, degree of crosslinking, and behaviour in the rubbery state, which will be briefly referred to.

Only some of the well-characterized polymers are dealt with based on results obtained during the last decade mainly in the author's laboratory at the Institute of Chemical Physics in Moscow.

The simple chemistry of curing, the homogeneous nature, and good properties of glasses based on the above-mentioned reactants, allow a better understanding of some important aspects of structure-properties relationships of these polymers as compared to more complicated epoxy systems. Many of these results seem to be generally valid and applicable to networks of different chemical nature.

## 2 Important Features of the Chemistry and Topology of Epoxy-Aromatic Amine Networks

### 2.1 Reactions in the Cure Process

Polymers to be considered below are networks formed in the following reaction:

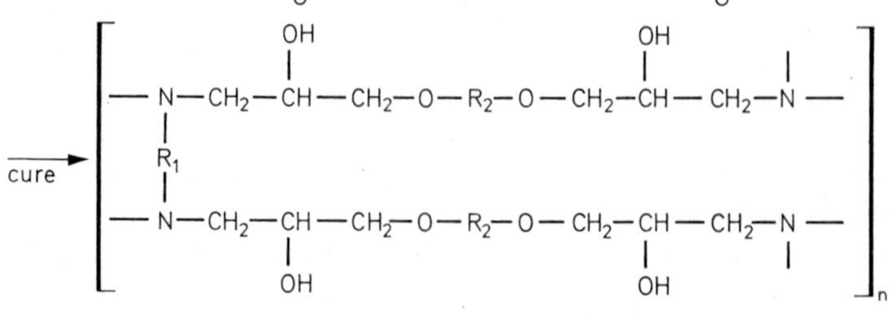

$$H_2N-R_1-NH_2 + H_2C\underset{O}{-}CH-CH_2-O-R_2-O-CH_2-CH\underset{O}{-}CH_2 \xrightarrow{cure}$$

$$\left[ \begin{array}{c} -N-CH_2-\underset{\underset{R_1}{|}}{CH}(OH)-CH_2-O-R_2-O-CH_2-\underset{|}{CH}(OH)-CH_2-N- \\ | \\ -N-CH_2-\underset{|}{CH}(OH)-CH_2-O-R_2-O-CH_2-\underset{|}{CH}(OH)-CH_2-N- \end{array} \right]_n \quad (I)$$

where: $R_1 =$ [structures: phenyl; dimethylpyridyl; diphenyl sulfone] and

$R_2 =$ [structures: phenyl; diphenyl isopropylidene (with CH$_3$/CH$_3$)]

In a few cases $R_2$ is: [phthalate-type structure]; [cyclohexane dicarbonyl]; [phenyl] and

$R_1 =$ [diphenylmethane structure with CH$_2$ bridge]

All the reactants are individual compounds of high chemical purity with epoxy equivalent close to theoretical values. It has been shown [6,7,16,17] that, at $T_{cure} \leq$ $\leq 110–120\,°C$ and especially with diglycidyl ether of resorcinol (DGER) or Bisphenol-A (DGEBA) and m-phenylenediamine (m-PhDA) as reactants, Eq. (I) proceeds practically without side reactions and yields a polymer with a chemical structure very close to that shown in above. Using different initial ratios of reactants ($P = [NH]_0/[EP]_0$ molar ratio), one can prepare a family of polymers with rather broad variations of crosslink concentration per unit volume or number of unreacted groups of different type and properties of the final products.

Using aniline (AN) or a mixture of AN with m-PhDA as a curing agent, one can prepare another set of polymers (from linear polymers to cured ones) with different concentrations of chemical crosslinks. Finally, the use of monofunctional epoxy compounds like phenylglycidyl ether (PhGE) allows to modify the network structure of the cured product. Thus, we can obtain a set of polymers from chemically similar structural blocks, combined in a different manner, with a rather broad spectrum of physical and mechanical properties. An important chemical factor affecting network formation is the relative reactivity of a primary and secondary amino group participating in the curing. If a bifunctional primary amino group is more reactive, one may expect the formation of long linear macromolecules at the beginning of the cure process. The next stage of network formation is the crosslinking between coiled linear polymeric chains. It is accepted that regions differing in crosslink density can be formed in a polymer under these conditions [13].

The question of relative reactivity of a amino groups in different systems is analyzed in detail by B. Rozenberg in this volume and in Ref. [7]. In the systems under consideration, relative reactivity of amino groups is rather close [6] and cannot be the reason for nonhomogeneous crosslink density.

To describe in detail the chemical structure of the networks formed in Scheme I, one has to know the reaction conversion $\alpha$ under all the reaction conditions. To measure α we used several methods: calorimetry (DSC [19,20]) and Calvet-type [16]), FTIR spectroscopy [21], and chemical titration of unreacted epoxy groups. All the methods gave very comparable results with deviations of the α values of about ±1% up to the latest stages of cure.

It is well known [6,8,13] that different side reactions may occur during the thermosetting of epoxides. This may lead to many differences in the chemical and physical structure of a final polymer. The main side reaction proceeding in a polymer under $T_{cure} > 130°–140\ °C$ [22] is the reaction between secondary OH groups and unreacted epoxy rings:

$$\sim\!\!\!N-CH_2-CH(OH)-CH_2-O-R_2 + H_2C\overset{O}{-}CH-CH_2-O-R_2 \longrightarrow$$

$$\sim\!\!\!N-CH_2-CH(O-CH_2-CH(OH)-CH_2-O-R_2\!\sim)-CH_2-O-R_2\!\sim \qquad (II)$$

The rate of this reaction is about ten times smaller than that of Eq. (I) and under typical cure conditions it becomes noticeable only after the end of the main process. Scheme (II) changes the structure of networks by the formation of additional crosslinks of the ether type. This makes the total connectivity of the network higher. This structural change influences some properties of polymers in the glassy and rubbery state (see Sects. 4 and 5), but it is really pronounced in nonstoichiometric systems with an excess ($P < 0.8$) of epoxy components [19,22]. Crosslinks of the ether type may principally appear in polymers due to a condensation reaction between OH groups. Under our conditions this process normally does not take place.

Furthermore, the chemical structure of networks are changed by thermal oxidation reactions [17,23,24]. These are rather important for epoxy networks with aliphatic amines since they usually take place in the presence of air at $T \geq 130\ °C$. In aromatic amine-based polymers this kind of reaction becomes important at $T \geq 220°–240\ °C$ [17,23]. The only exception are polymers with a large excess of epoxy groups in the initial mixture. For example, the polymer with $P = 0.4$ [23] starts loosing its weight at $160\ °C$ [17,23,24]. All polymers considered in this paper are prepared from mixtures with $0.6 \leq P \leq 1.6$. Cure and post-cure treatment temperatures are below $190\ °C$. This means we may not consider thermal oxidation processes in our structural analysis of the networks.

In some cases, network structure is modified by aminolysis reactions [25]. An example is the polymer formed from diglycidylic ester of o-phthalic acid and diaminodiphenilmethane. Aminolysis makes the chain between crosslinks shorter and influences the properties of the polymer (dynamic shear modulus in a rubbery state [25]). This process should be taken into account for ester-type epoxies at $T_{cure} \geq 70\text{--}80\ °C$.

In conclusion to the short analysis on curing chemistry of epoxy-aromatic amine networks (for more detailed analysis see papers of K. Dušek and B. Rozenberg in this volume), one can say that the chemical structure of the polymers under consideration is mainly determined by the curing reaction in Eq. (I). Equation II becomes important only for polymers with an excess of epoxy groups at $T \geq 150\ °C$. This rather simple situation makes the analyzed polymers very suitable for basic investigations.

## 2.2 Vitrification During Cure

Vitrification of a reacting system is a characteristic feature of a thermosetting process. Usually, cure proceeds when $T_{cure} = $ const. and vitrification appears as a result of molecular weight increase. Vitrification during cure takes place only when $T_{cure} < T_g^\infty$, where $T_g^\infty$ is the maximum glass transition temperature (Sect. 4) of the polymer with the main curing reaction of Eq. (I) completed and conversion reaching its topological limit of $\alpha^{top}$ (see Sect. 2.3). When $T_{cure} > T_g^\infty$, a completely cured network polymer exists in its rubbery state. For the cured epoxies discussed here, the maximum value of $T_g^\infty = 145\text{--}160\ °C$ (stoichiometric mixtures). This means that typical cure ($T_{cure} = 80\text{--}120\ °C$) processes of epoxy-aromatic amine mixtures proceed with a vitrification phenomenon. The important feature of vitrification is as follows: chemical reactions become diffusion-controlled and do not reach completion.

Figure 1 shows kinetic curves of the DGER-*m*PhDA (P = 1) cure reaction at different $T_{cure}$. From these curves, the existence of the conversion at which the reaction stops due to diffusion control, $\alpha^{dif}$, can be seen. The measurement of the total reaction heat Q at different $T_{cure}$ and its comparison with the estimated value based on the specific heat of epoxy ring opening give the values of $\alpha^{dif}$ for any $T_{cure}$ [16]).

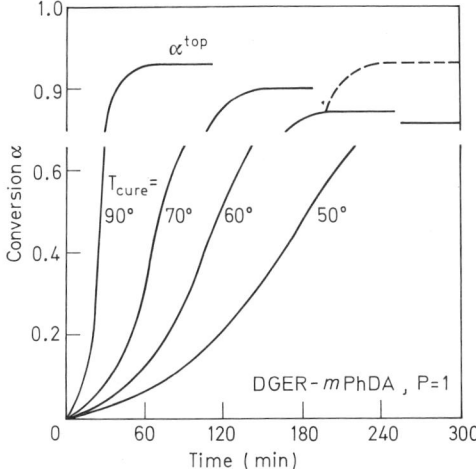

**Fig. 1.** Rate of isothermal cure of the DGER-*m*PhDA (P = 1) system at different $T_{cure}$. Dotted curve: after a jump change of $T_{cure}$ from 60 °C to 90 °C

Such a behaviour of epoxy systems has been known for a long time [7,27,28,29]. A detailed description of the situation in terms of TTT diagrams [5] is given in this volume (see paper by J. K. Gillham). Any rise of $T_{cure}$ after the reacted mixture has reached $\alpha^{dif}$ leads to the continuation of a thermosetting process up to the next diffusion limit, which corresponds to the next $T_{cure}$ (dotted curve in Fig. 1). This stepwise curing stops only when $T_{cure}$ becomes close to $T_g^\infty$ of the cured samples [15,19]. The vitrification during cure leads not only to kinetic complexities but also to changes in the structure and properties of cured polymers [15,20]. Systematic decrease of $T_{cure}$ finally gives polymers with very unusual properties of glassy state; this will be considered in Sect. 6. During cure, when vitrification sets in, especially in the early stages of the reaction (low $T_{cure}$), phase separation processes can occur. As was found in Ref. [30], unreacted epoxy reactants can form small crystals which appear in a cured polymer and influence some of its properties. We checked the situation in all polymers considered and found that such a kind of phase separation could appear only if DADPhS (4,4'-diaminodiphenyl sulfone) was used as curing agent and $T_{cure} < 70–80\ °C$. In all other cases, there was no evidence of phase separation or crystallization.

## 2.3 Network Topology

The description of a network structure is based on such parameters as: chemical crosslink density and functionality, average chain length between crosslinks and length distribution of these chains, concentration of elastically active chains and structural defects like unreacted ends and elastically inactive cycles. However, many properties of a network depend not only on the above-mentioned characteristics but also on the order of the chemical crosslink connection — the network topology. So, the complete description of a network structure should include all these parameters. It is difficult to measure many of these characteristics experimentally and we must have an appropriate theory which could describe all these structural parameters on the basis of a physical model of network formation. At present, there are only two types of theoretical approaches which can describe the growth of network structures up to late post-gel stages of cure. One is based on tree-like models as developed by Dušek [7,10,26]. The other uses computer-simulation of network structure on a lattice; this model was developed by Topolkaraev, Berlin, Oshmyan [9,31] (a review of the theoretical models may be found in Ref. [7] and in this volume by Dušek). Both approaches are statistical and correlate well with experiments [6,7,9,10,13,26,31]. They differ mainly mathematically. However, each of them emphasizes some different details of a network structure.

In our analysis and explanations of macro-properties of epoxy-aromatic amine networks, we mostly used the structural features which resulted from computer simulation [6,9,15,19]. In this section, we shall give a brief description of this model [9,13,19,31].

The basic idea of the model was the representation of a network as a statistical combination of molecular cycles with a different number of diepoxides and amines in each of them. This approach gives the possibility to describe network topology by a distribution function of cycle sizes or by its moments. Modelling also includes small cycles (monocycles), which appear as structural defects of the network — elastically inactive cycles. One can find some examples of different cycles in the paper by B. Rozenberg in this volume.

Network formation was generated on the lattice (hexagonal, tetragonal and cubic) with the total number of crosslinks amounting to $2-4 \times 10^2$. Tetrafunctional amine molecules were fixed in the lattice knots. This lattice was "immersed" in uniform liquid of diepoxide molecules which could penetrate the lattice freely. The network appeared as a result of an addition reaction between diepoxides and amine groups of crosslinks.

The order of addition was simulated by the Monte-Carlo procedure. Diepoxides, once connected with an amino group, can react only with the closest neighbours in the first coordination sphere.

This assumption leads to a very important conclusion about the existence of the topological conversion limit $\alpha^{top}$ in the completely cured system.

At the latest stage of the cure reaction in the stoichiometric network, there must be an equal amount of unreacted groups as well as epoxy groups having reacted with their neighbours, but having no closest neighbours for the second addition. Due to the high connectivity of the network, these spatially separated (isolated) unreacted groups cannot meet and react with each other under normal reaction conditions. Computer modelling gave $\alpha^{top} = 0.95$–$0.96$ [9, 16, 31] for DGER-$m$PhDa networks (P = 1). Experimental measurements showed $\alpha^{top} = 0.92$–$0.93$ which did not change up to $T_{cure} = 180\ °C$ for several hours [16, 19, 21].

Computer simulations were performed taking into account the component ratio P, the relative reactivity of primary and secondary amines, and different probabilities of monocycle formation. The simulations showed the structural features of networks (both topology and defects) at all stages of the cure process. Two examples of network structure received from computer simulation are shown in Fig. 2. Here

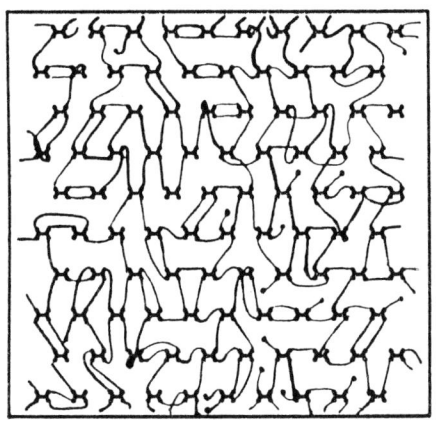
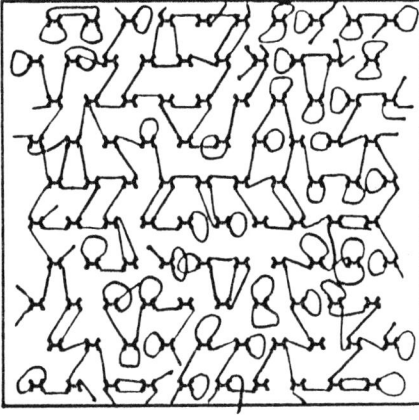

a        b

**Fig. 2a and b.** Two examples of network topological diagrams obtained by computer simulation. a) low and b) high probability of monocyclization. ① ⨞ — lattice crosslinks (aromatic amines); ② ⌒ — unreacted ends

one can clearly see the cycles of different size and network defects as isolated unreacted groups and monocycles.

The results yielded detailed information about chemical defects which appeared in each of the cured systems. We are especially interested in the structures like "defective lattice crosslinks"-amine molecules with a different number of links with the entire network.

Scheme (III) shows all possible cases of these knots for systems with *m*-PDA as the curing agent:

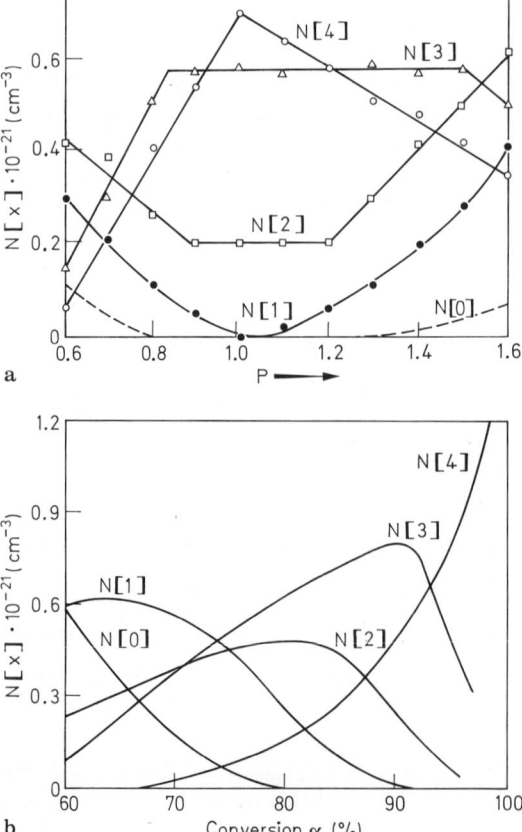

Fig. 3a and b. Calculated chemical crosslink concentrations N(0 ÷ 4) during cure b) and for completely cured networks at different P a)

where R is either a diepoxide molecule or H atom and ~| shows the connection of the knot with the entire network.

During simulations, one can follow the kinetics of formation of different crosslinks for the given "sample" or obtain the number of knots of different connectivity. Some of the results are given in Fig. 3 which shows the change in the concentration of different types of knots (from 0 to 4 linked) during cure (Fig. 3b).

At present, there are no physical methods to measure the concentration of amines of different types in networks and thus we cannot experimentally prove the computed values. However, the computed results seem reasonable since computer simulations give many features of the real behaviour of the systems under consideration. For example, the calculations gave kinetic curves of different reacting mixtures, sol and gel fractions, and equilibrium rubbery modulus. All results showed very good correlation with experiments [6,9,13,16,19,31]. This situation allows us to correlate the structural features of networks (for example, relative amounts of defects) obtained from computer simulations with macroscopic properties of the polymers.

## 2.4 Main Results of Network Topological Analysis

The topological computer model of epoxy-aromatic amine networks showed the following very important structural characteristics of polymers:

(i) Networks are topologically rather nonuniform. This is seen in Fig. 2. This nonuniformity appears even with monodisperse distribution of initial functionalities of reactants and in the absence of a monocyclization reaction. It is displayed through a broad cycle size distribution and the existence of crosslinks with different connectivities. However, this nonuniformity is not the result of cluster formation of some reactants or phase separation occurring during the reaction, since it was forbidden in the simulation. Thus, the nonuniformity is the result only of the reaction statistics.

(ii) For stoichiometric mixtures there exists a limit of the cure reaction. The process never reaches cure conversion $\alpha = 1$ but has a so-called conversion topological limit $\alpha^{top} < 1$. Such a situation appears to be due to the topological isolation of reactive groups during the statistical course of a thermosetting reaction.

(iii) The topological model gives the possibility to calculate the equilibrium elasticity modulus of networks in the rubbery state taking into account the topological nonuniformity of a polymer and the real elasticity of a chain between crosslink [10,33,34] (see Sect. 5).

The calculated equilibrium network moduli of elasticity and experimental moduli measured in the rubbery state [10,34] of different epoxy-amine networks show quantitative agreement. This means that the real structural nonuniformity of crosslinked epoxy-aromatic amine polymers exists only at the level of statistical deviations of network structure from the ideal one. A comparison of $E_{calc}$ and $E_{exp}^{\infty}$ is shown in Fig. 4.

The latter result shows that to interpret the mechanical properties of networks we do not need to take into account the spatial inhomogeneities of crosslink distribution in a sample, at least in the rubbery state. The analysis of epoxy networks performed under the framework of a tree-like model and experiments [7,10,26] brought the

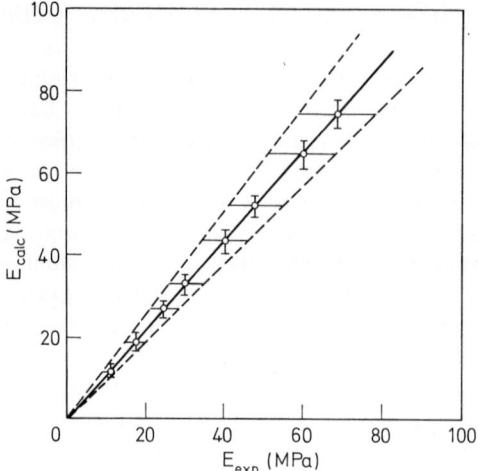

**Fig. 4.** Correlation between calculated Young moduli ($E_{calc} = e_i k_c N^{-1/3}$) and experimental values $E_{exp}$ in rubbery networks of DGER-DADPhS with different P

authors to practically the same conclusions about homogeneity and the nature of elasticity of networks. There are many approaches [30,35] where authors try to correlate the mechanical properties of epoxy networks in the glassy state with the spatial heterogeneity of crosslink density. However, the present results and the SAXS analysis of epoxy polymers [36,37] show that this correlation cannot play an important role in the systems under consideration.

## 3 Structural Features of the Glassy State

Under normal conditions, polymers formed in Scheme (I) are amorphous (X-ray, IR spectroscopy) with $T_g > 100\ °C$, high concentration of crosslinks per unit volume ($\sim 10^{21}\ cm^{-3}$) and about the same number of OH groups. All OH groups are involved in hydrogen bonding [6,21]. In this Section, we shall consider some properties of epoxy-amine networks in the glassy state.

### 3.1 Packing Density

The excess of volume is one of the most important features which distinguishes the amorphous from the crystalline state of matter of the same chemical structure. In an amorphous body, one can observe some excess volume which exists in the form of free volume and plays the key role in the kinetic properties of the body [38].

The structural perfectness of amorphous solids is usually characterized by the packing density K [39] which is determined by

$$K = \frac{\sum_i V_i}{M/\varrho}$$

where $\sum_i V_i$ is the volume occupied by the atoms of a polymer repeat unit. M is the molecular weight of the repeating unit, and $\varrho$ the density. For organic substances including linear polymers, the K values are very well documented [39,40], but data on networks are not available. Some authors believe that chemical crosslinks suppress molecular mobility of the network compared to a linear analog and introduce some geometrical constraints on the packing of atoms. The combination of these factors is bound to reduce the K values of the networks with a corresponding increase of the unoccupied volume fraction in a glassy network polymer. However, the measurement of the density of the polymers under consideration and the calculation of K [14,15,25] shows a very different picture. K values at 20 °C, $K_{20}$, for all epoxy-aromatic amine networks are $K_{20} = 0.705$–$0.728$ which is markedly higher than those ($K_{20} = 0.681 \pm 0.01$) of glasses of linear polymers.

Annealing and quenching of network glasses from above $T_g$ to the glassy state change the density not more than by 0.1–0.2%, which is the same as the change for linear polymers [41,42]. This result unequivocally shows that chemical crosslinks even of high concentration do not restrict any rearrangements important for the realization of the highest possible packing density. The appearance of crosslinks in a polymer only provides a better packing of all chain fragments. This situation shows that the mobility of a chain fragment of epoxy-aromatic amine networks is high enough even in the presence of chemical crosslinks. During the thermosetting reaction, the volume contraction for stoichiometric networks is about 8–9%. Evidently, such a high value is not due to rearrangements of chemical bonds during cure. The formation of new crosslinking chemical bonds may provide only about 2% of the total contraction. The cooling of the cured polymer from $T_{cure}$ down to room temperature may also provide 1%–1.5% contraction. This means that 5–6% of cure contraction is provided by better packing of network chains (both in the glassy and rubbery state) due to their higher connectivity as compared with low-molecular-weight compounds and linear polymers.

## 3.2 Heat Capacity $C_p$ and Thermal Expansion Coefficient

Figure 5 (a, b) shows typical heat capacities and thermal expansion coefficient curves for some epoxy-aromatic amine networks. Table 1 gives some numerical values for networks with different component ratios P: $C_p$ and β values for the glassy state do not practically depend on the chemical composition of networks and are very

**Table 1.** Thermal properties of DGER-$m$PhDA (P = 1) networks in the glassy and rubbery states

| $P = \dfrac{[NH]_0}{[EP]}$ | $T_g$, °C | $\beta_g$, $10^4$ K$^{-1}$ ±0.02 | $\beta_l$, $10^4$ K$^{-1}$ ±0.1 | $C_p$ (60°), J mol$^{-1}$ K$^{-1}$ ±5% | $\Delta\beta T_g$ |
|---|---|---|---|---|---|
| 0.6 | 68 | 1.86 | 4.8 | 1.68 | 0.144 |
| 0.8 | 108 | 1.86 | 4.5 | 1.44 | 0.120 |
| 1.0 | 144 | 1.87 | 4.4 | 1.4 | 0.124 |
| 1.3 | 128 | 1.88 | 4.6 | 1.44 | 1.129 |
| 1.5 | 114 | 1.90 | 4.8 | — | 0.138 |

[a] Subscripts g and l correspond to the glassy and liquid states respectively.

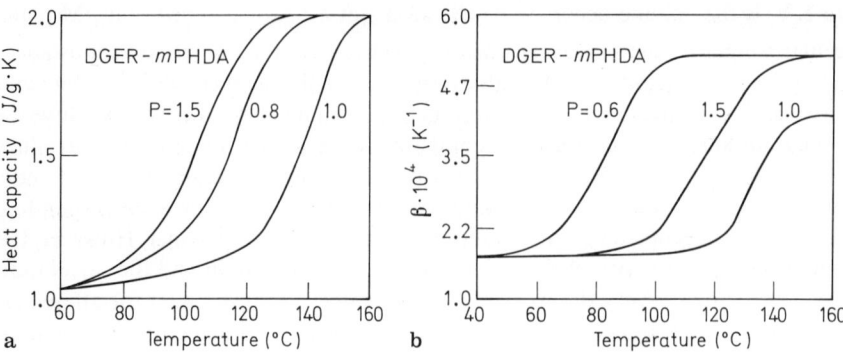

**Fig. 5a and b.** Heat capacities $C_p$ (**a**) and cubic thermal expansion coefficient $\beta$ (**b**) for DGER-*m*PhDA network of different P.T = 5–15 K/min

close to the data for various organic glasses. The same values for the rubbery state of networks are rather sensitive to their chemical composition and the concentration of crosslinks. This fact proves that the configurational mobility of a polymer depends on the chemical structure of the polymer. This is in contrast to the glassy state. Interesting information on the glassy state of networks and structural changes near $T_g$ is provided by the values of heat capacity $\Delta C_p(T_g)$ and thermal expansion $\Delta \beta(T_g)$ jumps. The values of the jumps as a function of P for DGER-*m*PhDA networks are given in Fig. 6 (a, b). $T_g$ and $\Delta C_p$ were measured by the methods described in Refs. [44, 19, 39]. Both values show a minimum for stoichiometric mixtures (P = 1) due to the highest density of crosslinks. This is evident from the comparison of $\Delta C_p(T_g)$ values of uncured and cured epoxy systems (Fig. 6a). Since the values of

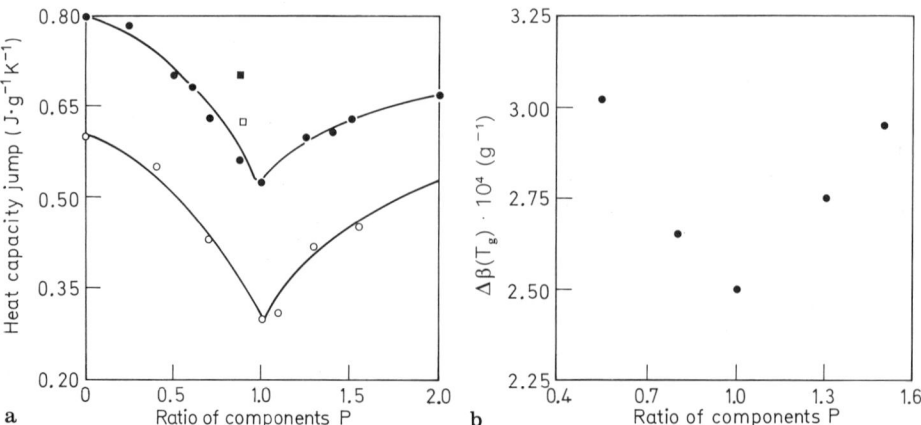

**Fig. 6a and b.** Heat capacity $\Delta C_p(T_g)$ and cubic thermal expansion coefficient jumps $\Delta \beta(T_g)$ for some epoxy-amine systems. Experimental results: ○ DGER-*m*PhDA networks of different P, completely cured and annealed (T = 0.5°/min); ● DGEBA-*m*PhDA networks of different P, completely cured and annealed; □ DGEBA-*m*PhDA unreacted mixture (P = 1); ■ DGER-*m*PhDA unreacted mixture (P = 1); solid lines — the best fit

jumps are evidently a function of network chemical composition, it is reasonable to rationalize them in terms of increments specific for each part of the composition. In this case, it is convenient to choose as structural elements the molecules of diepoxides, diamines and crosslinks — a crosslink in this case means a new chemical bond between reacted amine and epoxy groups. It is clear that the formation of every new chemical bond limits the configurational freedom of the mixture and decreases $\Delta C_p(T_g)$. One may write

$$\Delta C_p(T_g) = M_1 \Delta_1 + M_2 \Delta_2 + M_3 \Delta_3, \qquad (2)$$

where $M_1$, $M_2$ and $M_3$ are the numbers (moles) of diepoxides, diamines and crosslinks per gram of mixture and $\Delta_1$, $\Delta_2$, $\Delta_3$ are the increments in a heat capacity jump $\Delta C_p(T_g)$ per mole of the corresponding units. In Table 2 are given the values of $\Delta$ obtained from the best fit parameters of this additive Equation. Negative values of $\Delta_3$ show that crosslinking decreases the configurational mobility of the network polymer above its $T_g$. Using $\Delta$ values from Table 2, one can fit the experimental results (solid lines in Fig. 6a).

**Table 2.** Additivity parameters for the network heat capacity [45)]

| Network | $\Delta_1$ | $\Delta_2$ | $\Delta_3$ | $\delta^a$ |
|---|---|---|---|---|
| | J mol$^{-1}$ K$^{-1}$ | | | |
| DGER-mPhDA | 177 | 142 | −53 | 4% |
| DGEBA-mPhDA | 204 | 154 | −58 | 6% |

[a] Root-mean-square deviation of experimental values from the calculated ones.

**Table 3.** Parameters of the hole theory for some linear polymers and networks

| Polymer | $\Delta \beta T_g$ | $\Delta C_p T_g$ J·g$^{-1}$ | $N_h$ number of holes per gram at $T_g^\infty$ ($\times 10^{21}$) | $E_h$ hole energy (kJ/mole) | $F_g$ hole fraction | $V_g$ hole volume (Å$^3$) |
|---|---|---|---|---|---|---|
| Polyisobutylene | 0.091 | 80 | 1.9 | 6.1 | 0.024 | 20.8 |
| Polystyrene | 0.140 | 126 | 2.6 | 9.5 | 0.047 | 29.3 |
| Linear polymers[a] | 0.120 | 110 | 2.6 | 8.2 | 0.033 | 21.3 |
| DGER-mPhDA | | | | | | |
| P = 0.7 | 0.127 | 215 | 5.2 | 9.75 | 0.036 | 11.5 |
| P = 1.0 | 0.124 | 199 | 4.3 | 11.1 | 0,038 | 15.1 |
| P = 1.3 | 0.129 | 237 | 5.7 | 10.2 | 0.041 | 12.1 |

[a] calculated as average over 9 typical linear polymers.

Heat capacity and thermal expansion jumps are suitable parameters for comparison of glassy networks with linear polymers in terms of the hole model [46,47].

In Table 3 are presented some values characterizing the glassy state of several linear polymers and DGER-*m*PhDA networks. The values of $\Delta\beta T_g$ (Simha-Boyer factor) for networks are practically the same as for linear polymers. This means that the glass transition of these networks occurs when a free volume *fraction* becomes the same as that $T_g$ for glasses of linear polymer. On the contrary, $\Delta C_p T_g$ values of the considered networks are twice as high as those for linear polymers. The hole concentration $N_h$ and the energy of hole formation $E_h$ are also considerably higher for networks. From these results, it follows that the free volume necessary for network glass transition appears in the form of a larger number of smaller holes as compared with glasses of linear polymers.

Boyer [48] has introduced into polymer science the $\Delta C_p T_g$ quantity and has found its correlations with some properties of amorphous polymers. It is interesting to compare the data for epoxy-amine networks with those given in Boyer's paper. The

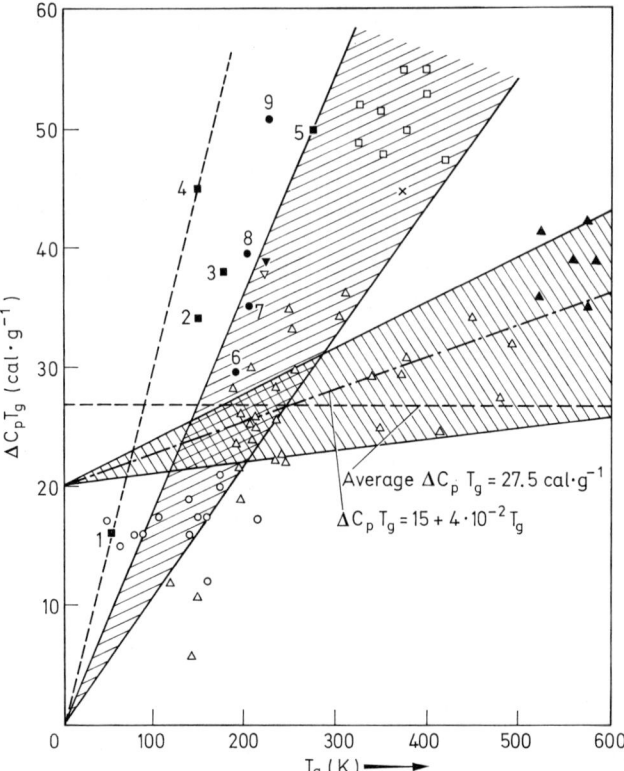

**Fig. 7.** Variation of $\Delta C_p T_g$ with $T_g$. △ Boyer's data [48]; ▲ polyquinoxalines [88]; ○ hydrocarbons and higher alcohols [48]; □ DGER-*m*PhDA networks, P = 0.6–1.8; ▼ DGER-*m*PhDA, (P = 1) unreacted mixture; ▽ DGER; × network DGER-*m*PhDA-AN(0.5), P = 1; ■ Alcohols: *n*-propanol (1); propylene glycol (2); glycerol (3); ethanol (4); glucose (5). ● Polyethers and polyacetals: (Poly(tetrahydrofurane) (6); polydioxolane (7); polytetramethyleneoxide (8); polyoxymethylene (9)

results are shown in Fig. 7. It is clear that the picture does not look as universal as it appeared originally [48]. It seems that the polymers for which the values $(\Delta C_p T_g)$ can be found in the literature can be divided into two groups. It is difficult to say now why epoxy systems have rather high $(\Delta C_p T_g)$ values, but it is interesting to note that the majority of the systems (low- and high-molecular-weight) with the largest $(\Delta C_p T_g)$ are all hydrogen bonded systems.

## 3.3 Hydrogen Bonds in Epoxy Networks

Hydrogen bonds are one of the most representative structural features of epoxy polymers. Each step of the cure reaction involving epoxy ring opening forms one OH group, whose concentration in a cured polymer is high.

Practically all OH groups are involved in H bonding in a bulk polymer [19], since there are many proton acceptors available in epoxy-aromatic amine networks. At room temperature the concentration of free OH groups in networks is lower than 1–2%. The formation enthalpies of different H bonds in the networks measured by the shift of $\nu(OH)$ vibrations in IR spectra are shown in Table 4. It is seen that the largest part of all H bonds ($\approx 90\%$ in a stoichiometric mixture) comes from the autoassociation of OH groups.

In non-stoichiometric mixtures with $P > 1$, the unreacted NH groups are involved in H bonding which distinguishes this kind of networks from those with $P < 1$.

To understand the role of H bonds in glass transition molecular rearrangements, it is interesting to follow the concentration of H bonds with temperature increase. In Fig. 8, the change in the concentration of H bonds (—OH ... OH type) with temperature is shown, which is determined from the change of $\nu(OH)$ band intensity in FTIR spectra [49]. It should be noted that there is no change in the slope of the concentration curve (Fig. 8) in the $T_g$ region. This means that it is difficult to relate structural rearrangements of the polymer near $T_g$ to an abrupt change of H-bonded groups in the network. The concentration of OH groups of the polymer involved in hydrogen bonding is still about 50% at 200 °C.

Some authors assign good mechanical properties of epoxies to high concentrations of H bonds. First, the Young modulus of a polymer is considered to be sensitive to the presence of H bonds. The Young modulus of a polymer can be described by:

$$E = \left(\frac{\partial^2 U}{\partial \varepsilon^2}\right)_T - T\left(\frac{\partial^2 S}{\partial \varepsilon^2}\right)_T, \qquad (3)$$

**Table 4.** Enthalpies of formation of H bonds for epoxy-aromatic amine networks

| $-\Delta H (25°)$, kJ/mol | Types of H bonds | | | | Networks with $P > 1$ | |
|---|---|---|---|---|---|---|
| | (OH ... OH) n | | OH ... O—$C_6H_5$ | | | |
| | $n \geq 3$ | $n = 2$ | OH ... N—$C_6H_5$ | | NH ... N | NH ... OH |
| | 18–19.5 | 12.5–14.5 | 8–10 | | 7–8 | 11.8–12.4 |

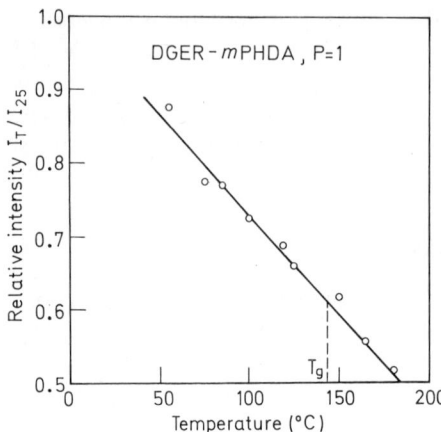

Fig. 8. Change of the relative amount of hydrogen bonded ν(OH) vibrations with temperature. $I_T$, $I_{25}$ — intensity at T and 25 °C. $I_T = (S_{3400}/S_{2930})_T$; S — integral intensity of the bands at $\nu = 3400$ cm$^{-1}$ and $\nu = 2930$ cm$^{-1}$

where ε is the strain, S the entropy and U the internal energy. In the glassy state, the Young modulus of polymers is determined by the first term of Eq. (3). H bonding increases the average atom—atom interaction (mean field). That is why it is reasonable to expect an increase of Young moduli. A rough estimate of cohesion energy increase of epoxy polymers due to H bonds gives a value of about 30–40%.

However, experimental results do not support these expectations. Young moduli of all the polymers considered here are rather close to those of organic glasses (polymeric and non-polymeric) [40, 62] (see also Sect. 5) at 25 °C, $E_{25} = 3.0$–3.7 GPa. These values show that the U term, even in densely crosslinked epoxy network glasses, has an intermolecular origin and is determined by Van-der-Waals interactions. Also the deformation of H bonds and the deformation of chemical bonds and valence angles of network chains do not play any role in the elasticity of glassy networks.

In Sect. 6, it will be shown that epoxy networks, cured at low $T_{cure}$, usually show markedly higher moduli although the concentration of OH groups is lower. Thus, it may be concluded that the high concentration of H bonds in epoxy polymeric systems is likely to influence the chemistry of network formation (due to the complexes of reagents through H bonding [6]) and configurational entropy in the glass transition region (change of $\Delta C_p T_g$, for example), but does not markedly influence such properties as $T_g$ and glassy modulus.

## 3.4 Local Molecular Motions

### 3.4.1 Thermally Stimulated Discharge (TSD) Measurements

Local molecular motions are the only motions in the glassy state and their importance is obvious in crosslinked glasses.

There is a number of literature data on the motions of γ and β types in glassy polymers [50]. However, it is not clear whether this picture is applicable to densely crosslinked polymers. The study of local motions in the networks under consideration has been performed using mainly the TSD technique [51] because of its high resolution and sensitivity.

In Fig. 9 (a–c), TSD traces for some epoxy-aromatic amine networks in the glassy state are shown. It is clearly seen that in the glassy state there exist several different types, at least four, of TSD peaks (Fig. 9c). A special measurement has shown that all TSD peaks are a result of dipole relaxation but not of the migration of ions or heterocurrents. From Fig. 9 a and b it becomes clear that the $\gamma_1$ peak belongs to unreacted epoxy groups; the $\beta$ peaks, undoubtedly, are related to some motions of aliphatic chains. The relaxation time as a function of temperature has been measured for two peaks — $\gamma_1$ and $\beta_2$ (dotted lines in Fig. 9c). Both peaks have been refined by

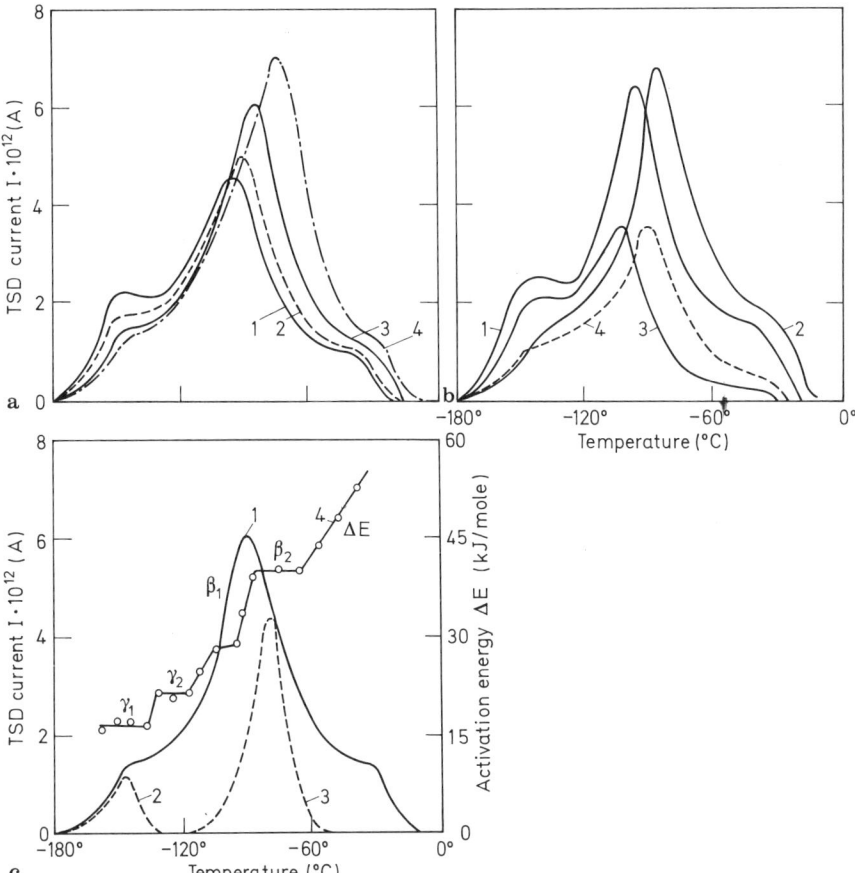

Fig. 9a–c. Thermally stimulated discharge (TSD) current as a function of temperature for some epoxy-aromatic amine networks. a) DGER-mPhDA, P = 1. Cure conditions: 1 $T_{cure}$ = 23 °C, 100 days; 2 $T_{cure}$ = 55 °C, 60 h; 3 $T_{cure}$ = 90 °C, 20 h; 4 $T_{cure}$ = 130 °C, 5 h. b) TSD curves for the networks: 1 DGER-mPhDA, P = 0.73; 2 the same, P = 1.47; 3 DGEBA-mPhDA, P = 0.7; 4 the same, P = 1.31. c) TSD curves (1, 2, 3) and activation energy (4) for completely cured DGER-mPhDA (P = 1) network. 1 Total TSD curve; 2, 3 individual $\gamma_1$ (2) and $\beta_2$ (3) peaks, separated from the total TSD curve (1) by peak cleaning procedure; 4 activation energy $\Delta E$ as a function of temperature. $\Delta E$ was measured by partial heating procedure [51, 52]. Polarization field $E_{pol}$ = 15 kV/cm. T = 5 °C/min

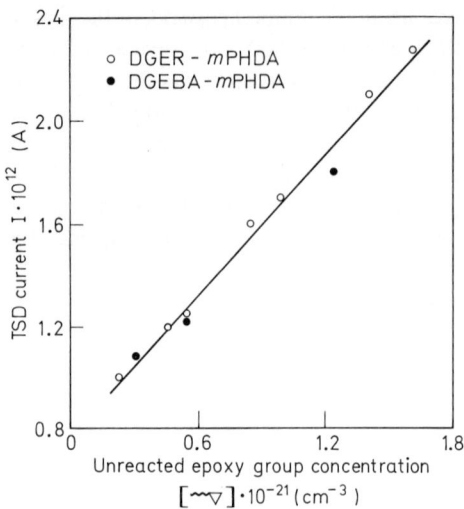

Fig. 10. TSD current at $T_{\gamma_1} = -148$ °C as a function of network unreacted epoxy-group concentration

"peak cleaning" and "partial heating" procedures [52, 51]. Both types of motions show Arrhenius-type behaviour with activation energies of 5 and 11 kcal/mole of mobile units. These results allow us to make some assumptions about the nature of the $\gamma_1$ and $\beta$ types of motions.

$\gamma_1$ *Peak*. In Fig. 10, the intensity of the $\gamma_1$ peak is given as a function of the number of unreacted epoxy groups in the polymer. The $\gamma_1$ peak corresponds to some rearrangements in the fragment >N—CH$_2$CH(OH)CH$_2$OR$_2$OCH$_2$CHCH$_2$. By using:

$$P_0 = \frac{N\bar{\mu}^2 E_{pol}}{3 kT_{pol}} \quad (4)$$

where N is the concentration of dipoles, $\bar{\mu}$ the dipole moment, $E_{pol}$ and $T_{pol}$, respectively, the field (kV/cm) and temperature of the sample polarization; the experimental saturation polarization value $P_0^{exp} = 1.3 \times 10^{10}$ c/cm$^2$ for the $\gamma_1$ peak one can calculate $\bar{\mu}$ from experimental data and compare it with the dipole moment $\bar{\mu}$ of the molecular model of the moving fragment. Table 5 shows the values of $\bar{\mu}$ and the corresponding $P_0^{calc}$ for some models of $\gamma_1$ moving fragments. From Table 5 it is clear that the only type of motion, which agrees with experimental values of $P_0^{exp} = 1.3 \times 10^{-10}$ c/cm$^2$ and dipole moment $\bar{\mu} = 1.2$ debye calculated from $P_0^{exp}$ using Eq. (4), is the epoxy ring rotation of the ⇌CH—CH$_2$ type.

$\beta$ *Peak*. Very often, the $\beta$ peak is referred to a "crankshaft" motion, which for this kind of networks, corresponds to [⇌O—CH$_2$—CH(OH)—CH$_2$⇌] fragments (dipole moment $\bar{\mu} = 1$ debye). Since the concentration of these fragments is known for networks of different composition, it is easy to calculate $P_0^{calc}$ by using Eq. (4). However, the calculation of $P_0^{exp}$ by integrating $\beta$ peaks of the experimental

**Table 5.** Average dipole moments $\bar{\mu}$ and saturation polarization values $P_0$ for molecular fragments of unreacted epoxy end groups in networks

| Fragment of structure | $\mu$, Debye | $P_0 \times 10^{10}$, coulomb/cm² |
|---|---|---|
| $>$N–CH$_2$–CH(OH)–CH$_2$–O–R–O–CH$_2$–CH(–O–)CH$_2$ | 2.5 | 5.4 |
| –O–CH$_2$–CH(–O–)CH$_2$ | 2 | 3.6 |
| (O-R-O diepoxide fragment) | 0.7 | 0.4 |
| –CH(–O–)CH$_2$ | 1.3 | 1.5 |

TSD curve gives a value of $P_0^{exp}$ which is higher by a factor of 2 than that calculated from the "crankshaft" structure.

On the other hand, the intensity of TSD β peaks is proportional to the volume concentration of ether groups in the polymer. "Peak cleaning" and "partial heating" procedures allow us to separate the $\beta_2$ process from the total TSD spectrum. Probably, one of the β peaks corresponds to crankshaft-type motion.

$\gamma_2$ *Peak*. The $\gamma_2$ peak is a process of unknown nature. Its activation energy is 6.5 kcal/mole (measured by "partial heating"). It possibly corresponds to the conformational transition of aromatic rings, which appear in epoxy polymers [53, 54] although the activation energy of the latter is higher ($\approx$ 10.5 kcal/mole). The recent data on phenyl-type motions in solid polymers give a value of 6.5 kcal/mol [54]. Realistic assignment of the $\gamma_2$ and $\beta_1$ peaks requires additional experimental investigations.

### 3.4.2 Conformations of Aliphatic Chains

Network chains between chemical crosslinks are evidently the only flexible elements of a glassy polymer and are involved in all types of molecular motion. As far as the local motions are concerned, the conformational changes in aliphatic fragments release this motion. It is important to know whether in densely crosslinked polymers there exists any restrictions concerning the conformational behaviour of these chains due to chemical crosslinks. Activation barriers and the conformational energy of chains can be changed by chemical crosslinking and influence the macroscopic behaviour of a polymer both in the glassy and rubbery states. Often, the differences in the properties of linear and crosslinked polymers are referred to the conformational behaviour of network chains. The conformations of aliphatic chains in epoxy-amine networks depend on the rotation about five skeleton bonds:

The conformations, can be detected by the changes in the intensity of sensitive bands in IR spectra by using model compounds (low-molecular-weight and linear polymers) and compare the results with networks. The differences in the behaviour of conformationally sensitive bands reflect the differences in the behaviour of aliphatic chains in different compounds. To detect the bands sensitive to the internal rotation, we have measured the spectra (50–4000 cm$^{-1}$) of model compounds A and B.

$$\text{Ph–O–CH}_2\text{–CH(OH)–CH}_2\text{–O–Ph–O–CH}_2\text{–CH(OH)–CH}_2\text{–O–Ph} \quad (A)$$

$$\text{(B): Ph–O–CH}_2\text{–CH(OH)–CH}_2\text{–N(–Ph–N–)–CH}_2\text{–CH(OH)–CH}_2\text{–O–Ph (tetrafunctional mPhDA–based structure)} \quad (B)$$

Which have in their structure aliphatic chains of the desired type. Both compounds are crystalline and their spectra give us the picture of the most stable conformations in the aliphatic chains. Spectral measurements of A and B in melt or $CCl_4$ solutions yield new bands which, undoubtedly, belong to the new conformers of the aliphatic fragments.

Table 6 gives the changes in vibrational band intensities due to melting [19] or dissolution.

The crystal-melt transition of compound A leads to a strong conformational change in aliphatic chains. The relative number of "non crystalline" conformers grows by ~40–50%.

It is important to note that all conformationally sensitive bands exist in the linear model polymer DGER-AN (P = 1) and in the DGER-mPhDA (P = 1) network [19]. The measurements of different FTIR spectra for network-linear polymer and linear polymer-A, B model compound pairs give a very important result: the content of different conformations in the aliphatic chains in network, linear polymer and low-molecular-weight model compounds is practically the same in the glassy state for all systems. Even the relative intensities of the conformational bands remain unchanged during different thermal prehistories (quenching or annealing) of the samples. The spectral picture for all compounds is the same in rubbery and glassy networks. This means that the crosslinks do not markedly influence the thermodynamic flexibility of the aliphatic chains of epoxy-aromatic amine networks.

The analysis of local motions shows that the motions, including conformational changes in aliphatic fragments of networks, have the same features — Arrhenius-type behaviour, the activation energies and $T_\beta$ values for linear polymers and low-molecular-weight A, B model compounds being equal. This fact strongly supports the idea that the internal rotation barriers in the considered fragments local motions

**Table 6.** Conformational-sensitive IR bonds in networks and model compounds [19]

| $\nu$ cm$^{-1}$ | | Intensity change in the crystal-liquid transition | Vibration Assignment | Possible Conformations | Internal Rotation around bond |
|---|---|---|---|---|---|
| crystal | liquid (melt, solution) | | | | |
| 1189 | 1191 | Decrease | $\nu(C_{arom}-O)$ | T | 1 |
| 1157 | 1157 | Increase | $\nu(C_{arom}-O)$ | C, G | 1 |
| 1471 | 1471 | Increase | $\delta(CH_2)$ | T | 2–5 |
| 1456 | 1458 | Decrease | $\delta(CH_2)$ | | 2–5 |
| 1388 | 1388 | Increase | $\gamma_w(CH_2)$ | T | 2–5 |
| 1361 | 1365 | Decrease | $\gamma_w(CH_2)$ | G | 2–5 |
| 1048/1038 | 1045 | Decrease | $\nu(C-O)$ | T | 2 |
| – | 1025 | Increase | $\nu(C-O)$ | G | 2 |

are not sensitive to the presence of crosslinks in the polymer structure. The same relative intensities of conformationally sensitive bands in the glassy and rubbery states of the networks probably mean that the flexible fragments of these polymers even in the glassy state (down to β transition temperature) have an equilibrium conformational structure. In this respect, the situation in densely crosslinked epoxy-aromatic amine glasses is similar to that in flexible polymers in the glassy state [55, 56] where the chains exist in the form of an unperturbed coil. This finding qualitatively explains some features of the epoxy-aromatic amine networks behaviour: high packing density is clearly determined by aromatic fragments of polymer, but very flexible aliphatic chains make the realization of the highest possible packing density easier. The packing density of the *m*-PhDA crystal at 25 °C is K = 0.84. It follows that the mechanical rigidity of the considered networks and their $T_g$ values are determined by packing of aromatic rings, primarily by the rings of the curing agent. The role of aromatic rings of amine in the glass transition behaviour of networks will be considered in Sect. 4.

## 4 Glass Transition Temperature

The glass transition temperature $T_g$ is one of the most important structural and technical characteristics of amorphous solids. The correlations of $T_g$ of linear polymers with their chemical composition, molecular weight, rigidity and symmetry of chains, as well as some other characteristics of macromolecules are well documented [57, 58]. The information on networks is much poorer. At present, for networks there exists mainly one parameter in structure-$T_g$ correlations. It is the concentration of crosslinks — a parameter which is very insufficient, since in networks there are chemical crosslinks of different functionality (connectivity) which are distinguished by their molecular mobility. This means that the topological aspect of the network structure should be taken into account in the $T_g$ analysis. Another difficulty connected with $T_g$ determination of polymers lies in vitrification occurring during polymer formation (Sect. 6).

When an isothermally cured system reaches the diffusion conversion limit $\alpha^{dif}$ corresponding to the chosen $T_{cure}$, all kinds of molecular motions responsible for the mobility of reacting groups become frozen. In physical terms, this means that the free volume fraction of the formed bulk polymer drops below a critical value. It is clear that the "frozen in" molecular motions are not a local type, since all motions on a smaller scale ($\gamma$ and $\beta$ types) exist unlimited in the glassy state at usual $T_{cure}$. Any increase of $T_{cure}$ after reaching $\alpha^{dif}$ (Fig. 1) is followed by a cure reaction. After this $T_{cure}$ rise, some large-scale molecular motions become again possible amd allow the reacting groups to establish mutual contacts and to react with each other.

The motions, which the system needs for the following cure, are to be of a segmental type. Thus, the given $T_{cure}$ must be close to the $T_g$ of the polymer cured up to $\alpha^{dif}$ at this $T_{cure}$. We shall call the glass transition temperature of this kind $T_g^{exp}$ (experimental $T_g$) in contrast to $T_g^{\infty}$ (glass transition temperature of the polymer cured up to its topological limit $\alpha^{top}$). The values of both $T_g^{\infty}$ and $T_g^{exp}$ may differ markedly. For the epoxy polymers under consideration they may differ by several tens of K [20]. The following relations hold: $T_{cure} \approx T_g^{exp}$ for $T_{cure} \leq T_g^{\infty}$; $T_g^{exp} \leq T_g^{\infty}$; with increasing $T_{cure}$, $\lim T_g^{exp} = T_g^{\infty}$.[1]

It is clear that $T_g^{exp}$ is a function of curing conversion or molecular weight (for linear polymers) at $\alpha^{diff}$. One can observe a noticeable difference between $T_g^{exp}$ and $T_g^{\infty}$ for such processes of polymer synthesis as polyaddition or condensation polymerization reactions. It is especially important for polymers with high $T_g^{\infty}$. For many heat-resistant polymers, $T_g^{\infty}$ is higher than the temperature limit of their chemical decomposition. We can never reach natural $T_g^{\infty}$ for these polymers. For such polymers, one really measures only $T_g^{exp}$, the value of which depends on the reaction conditions. For structure-glass transition temperature correlations of networks, $T_g^{\infty}$ is the most important quantity.

In this section, we shall consider only $T_g^{\infty}$ of epoxy-aromatic amine networks as well as the influence of side reactions on $T_g^{\infty}$. The magnitude of $T_g^{exp}$ will be analyzed in Sect. 6.

For $T_g^{\infty}$ measurements, calorimetry (DSC-990 Du-Pont), linear and volume cubic thermal expansion (TMS-2 Perkin-Elmer), thermo-stimulated discharge (TSD) technique, dynamic mechanical Young modulus and damping (DMA-981 DuPont) and dielectric ($10^2$–$10^5$ Hz) measurements have been used as experimental techniques. All methods gave very well correlated values of $T_g^{\infty}$ [19].

## 4.1 $T_g^{\infty}$ and the Chemical Structure of Networks

$T_g^{\infty}$ values of the networks prepared from different components of the initial mixtures are shown in Fig. 11, as measured by different techniques. The $T_g^{\infty}$ values reflect the chemical composition of the networks. $T_g^{\infty}$ decreases linearly with the content of unreacted amine and epoxy groups. Unreacted groups in networks play the same role as end groups in linear polymers. The stronger influence of epoxy groups

---

[1] These relations are correct without taking into account side reactions, which can break these relations and make $T_g^{exp} > T_g^{\infty}$. We shall consider this case in Sects. 6 and 4.2.

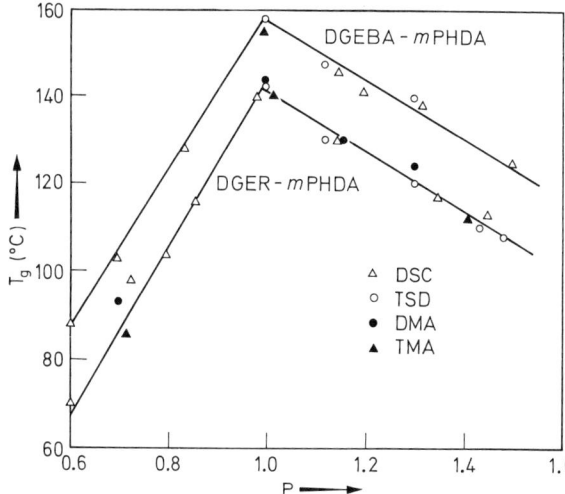

Fig. 11. $T_g^\infty$ values as a function of P

on $T_g^\infty$ can be explained by their higher molecular volume. However, the following question arises: which fragments in the chemical structure of epoxy-aromatic amine networks play the most important role in the motion determining the structural transformation into a glass? In other words, which fragment is to be changed chemically in order for $T_g^\infty$ of the epoxy networks to be changed? From the previous analysis (see Sect. 3) it is clear that these $T_g$-sensitive units cannot be fragments with hydrogen bonds and aliphatic chains. Even some aromatic rings cannot exhibit such an effect. Evidently, the most immobile fragments of networks are the crosslinks of the highest connectivity, namely amine molecules which are engaged four times. The comparison of the curve shapes in Figs. 11 and 3b supports this point of view. Only the concentration of N(4) — four times bound amines (Fig. 3b) — changes in the same manner as does $T_g^\infty$ (Fig. 11) for networks of different composition.

The highest $T_g^\infty$ is observed in stoichiometric mixtures with the highest concentration of N(4)-type crosslinks. The slope of the decrease of the concentration of these fragments is even higher for mixtures with an excess of epoxies (Fig. 11). To make the situation more representative, one plots $T_g^\infty$ as a function of the concentration of the N(4)-type crosslinks (Fig. 12). The concentration of N(4)-type amines in the polymers can be changed in different ways: by the change of the component ratio P (range of N(4) $0.2$–$0.7 \times 10^{-21}$ cm$^{-3}$ in Fig. 12) by preparing linear polymer DGER-AN (N(4) = 0) and networks DGER-AN-mPhDA with different ratios of (AN/mPhDA) but with stoichiometric ratios of epoxide and amine groups (N(4) < $< 0.4 \times$ qp$^{-21}$ cm$^{-3}$ in Fig. 12) and, finally, (N(4) > $0.5 \times 10^{-21}$ cm$^{-3}$) by high-temperature, long-time post-cure in which the originally unreacted epoxy groups are involved in side reactions (see Sect. 2) at temperatures of 170–180 °C. The linear correlation of $T_g^\infty$ with N(4) in a broad range of concentration of N(4) crosslinks supports the assumption that these crosslinks play the key role in the glass formation of epoxies. Recent NMR results [60] also support this idea. Not every aromatic group

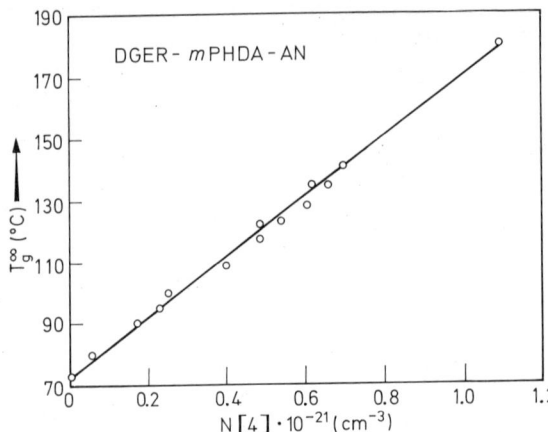

Fig. 12. $T_g^\infty$ as a function of N(4)-type crosslinks

influences $T_g^\infty$ behaviour of the networks. A comparison of $T_g^\infty$ of the networks having some unreacted groups with those in which these groups were replaced by monofunctional phenylglycidilic ether shows that the introduction of aromatic rings of type ⎯⟨◯⟩ does not change $T_g^\infty$ [19].

The dependence of the mobility of crosslinks on their connectivity (or functionality) influences some other features of α transitions in networks as compared with linear polymers.

In Fig. 13, the values of $T_g^\infty$ and apparent activation energy $\Delta E_{app}$ of α-transition [19] in networks are given $T_g^\infty$ increases by passing from linear DGER-AN polymers to more crosslinked networks of DGER-AN-mPhDA or DGER-mPhDA

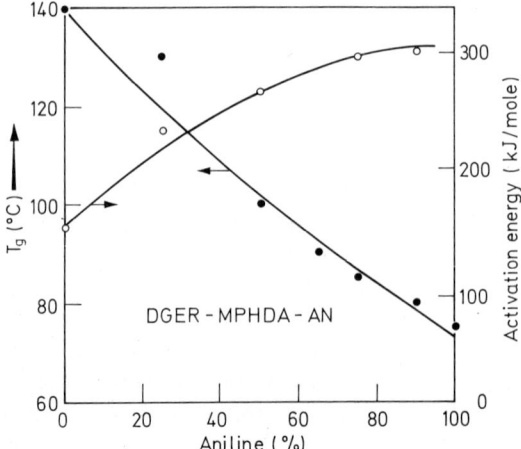

Fig. 13. Changes of $T_g^\infty$ and apparent activation energy $\Delta E_{app}$ of α transition with the concentration AN fragments in DGER-mPhDA-AN (P = 1)

(P = 1) systems while the apparent activation energy markedly decreases in the same series; $\Delta E_{app}$ for the linear polymer DGER-AN is close to the typical value for many polymers [57-59]. The results testify that the fragment (segment) of the network which is responsible for α-transition is in average smaller than that of a linear polymer. The temperature range of the $T_g^\infty$ transition for the networks is markedly narrower than that for linear polymers and epoxides prepared from industrial resins [13, 19]. This points to a narrower distribution of relaxation times in the considered networks as compared with other crosslinked polymers.

## 4.2 Glass Transition Temperature and Chemical Side Reactions

The side reactions (Sect. 2) can raise $T_g^{exp}$ so that $T_g^{exp}$ can be higher than $T_g^\infty$. However, the behaviour of networks in this respect depends on their composition. The high temperature post-cure treatments (T > 160 °C, 5–10 hours) of networks cured to $\alpha^{top}$ show that $T_g^\infty$ rises only in networks with an excess of epoxy components. The $T_g^\infty$ of cured stoichiometric and amine-excessive polymers does not practically change even after treatment at 180–200 °C for several hours. Therefore, that main reason for the $T_g^\infty$ change is the reaction between free epoxy and hydroxy groups, but not oxidation or degradation. The highest possible $T_g^{exp}$ value for the given sample can be reached only after a complete reaction of epoxy groups. The limits of $T_g^{exp}$ for the networks are given in Fig. 12. The polymers with $N(4) > 0.5 \times 10^{-21}$ cm$^{-3}$ were prepared in this way. Heating of the DGER-mPhDA (P = 0.66) network at 200 °C for 40 h gives with $T_g^\infty = 180$ °C [22].

The rise of $T_g^{exp}$ as a result of chemical side reactions is not desirable with respect to the material's properties. Usually, after such post-cure, polymers have higher Young moduli but become more brittle. The brittleness seriously lowers the applicability of these polymers as engineering plastics and matrices of advanced composites.

# 5 Mechanical Properties

## 5.1 Mechanical Properties in the Rubbery State

### 5.1.1 Experimental Results

Usually the deformation behaviour of rubbery crosslinked polymers with long chains between crosslinks obeys the classical theory of rubber elasticity. However, the situation for densely crosslinked polymers is not so simple.

In Ref. [26, 34] the mechanical behaviour of several epoxy-aromatic amine networks was analyzed in the rubbery state at temperatures of about $T_g^\infty + 40$ °C. Figure 14 gives some experimental results. If the rubber elasticity theory is obeyed, the following relation holds for uniaxial deformation:

$$\sigma = G_\infty(1 - 1/\lambda^2)$$

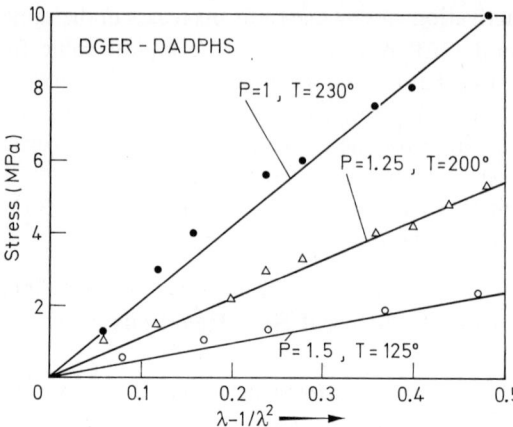

Fig. 14. Deformational behaviour of networks in the rubbery state. Solid lines correspond to the relation $\sigma = G_\infty(\lambda - 1/\lambda^2)$ with $G_\infty$ measured by torsion pendulum[34]

where $\lambda = 1 + \varepsilon$, $\varepsilon$ is the relative elongation and $G_\infty$ the equilibrium shear modulus. The linearity of the curves in Fig. 14 shows that the networks under consideration behave as an isotropic incompressible body. Direct measurements of $G_\infty$ by torsion pendulum [34] in the rubbery plateau gave $G_\infty$ values from 230 to 55 kg/cm² (Ref. [34]) for networks of DGER-DADPhS of different P. Other networks (DGER-mPhDA, DGEBA-mPhDA and DGER-DAP) show the same deformational behaviour. Two conclusions follow from the stress-strain behaviour of rubbery epoxy-amine networks:

— Densely crosslinked epoxy-aromatic amine networks behave qualitatively as typical rubbers or linear polymers of high molecular weight at temperatures above their $T_g^\infty$ [26, 34].
— Elongation at break, $\varepsilon_b$, of these polymers in the rubbery state is usually in the range of 12–35 % [12, 15, 34].

A relatively high $\varepsilon_b$ [12, 30, 35] in the glassy state is possible and it is not related to the inhomogeneous structure of the polymers. As it will be shown in Sect. 5.1.2., the largest possible elongation of intercrosslinked chains of these polymers is about 50 %.

5.1.2 Conformational Elasticity of Diepoxide Chains [33]

From the chemical structure of epoxy-aromatic amine networks, one can see that the diepoxide chain is the main elastic element of these polymers. However, the correctness of using Gaussian or Langevin approaches for the analysis of the experimental stress-strain data is uncertain. In the rubbery state, the equilibrium elasticity is determined by the conformational states of diepoxide chains. From the chemical structure of the chain and the geometry and energy of conformers, one can calculate the distribution function for the distance between chain ends, W(r), and consequently the equilibrium elasticity [61]. For diepoxide chains considered in the present paper, this calculation was performed by modelling diepoxide chain configurations using computer simulations (Monte-Carlo method) [33]. In these calculations, the analogy between semi-random walks with the step of chemical bond length was used. The correlations due to bond angles and excluded volume were taken into account. The main result of the simulation is the following: for short diepoxide

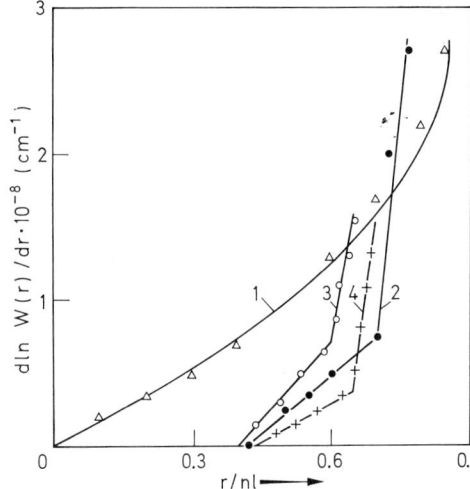

**Fig. 15.** Calculated values of d ln W(r)/dr as a function of relative elongation r/nl for DGER molecules. 1 Langevin approach; 2–4 Monte-Carlo analysis with energy difference between trans- and gauche-isomers: $E_t = E_g$ (2); $E_t - E_g = 0.5$ kcal/mol (3); $E_t - E_g = -0.5$ kcal/mol

chains the dependence of rigidity $kc = -kT(\partial^2 \ln W(r)/dr^2)$ on elongation has a step-wise character (Fig. 15). For the end-to-end distance $r \leq 0.8 \langle r_0^2 \rangle^{1/2}$, the rigidity is very close to zero but at $0.8 \langle r_0^2 \rangle^{1/2} < r < 1.45 \langle r_0^2 \rangle^{1/2}$ it becomes constant and equal to the rigidity of the freely-jointed chain calculated using the Langevin approach. At r/nl > 0.7 (l is the length of a chemical bond and n the number of bonds in the chain), the rigidity of chains becomes rather high (Fig. 16) due to finite chain extensibility. A variation of rotational energies of conformers does not qualitatively change the results.

### 5.1.3 Calculation of the Elasticity Modulus

It is well known that the elasticity of polymer networks with constrained chains in the rubbery state is proportional to the number of elastically active chains. The statistical (topological) model of epoxy-aromatic amine networks (see Sect. 2) allows to calculate the number of elastically active chains[1] and finally the equilibrium modulus of elasticity $E^{calc}$ for a network of given topological structure [9,10]. The following Equation [9] was used for the calculations of $E_\infty^{calc}$:

$$E_\infty^{calc} = e_t k_e N^{-1/3} \tag{5}$$

where $k_c$ is the rigidity of the network chain, $e_t$ is the effective modulus of a given topological network structure at $k_c = 1$, and $N^{-1/3}$ is the average distance between crosslinks. Equation (5) was obtained assuming that the main part of the network chains was deformed not far from their equilibrium end-to-end distances, i.e. $r \approx \langle r_0^2 \rangle^{1/2}$. The stress was applied to the borders of the topological network diagramme (see Fig. 2) and the deformation of the model network and stress distribution

---

[1] This number and the equilibrium modulus were calculated in Ref.[10] in the framework of a tree-like model. The results showed good agreement with the experiment [26].

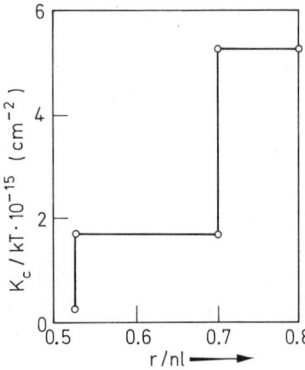

**Fig. 16.** The rigidity $K_c = -kT(d^2 \ln W(r)/dr^2)$ for a DGER linear chain in dependence on relative elongation $r/nl$ [33]

function through the network chains [9, 22] was calculated; $e_t$ is defined by the topology of the network.

The relationship between $E_\infty^{calc}$ and $E_\infty^{exp}$ is given in Fig. 4. A very good correlation shows that all chemical heterogeneities of the epoxy-aromatic amine networks under consideration exist in the rubbery state only on the level of statistical deviations (see also Ref. [10, 26]).

## 5.2 Mechanical Properties of the Glassy State

The transition of the polymer from the rubbery to the glassy state principally changes the deformation behaviour and mechanical response of the material. The dominant role of intermolecular forces in the glassy state of polymer fully suppresses the effect of the conformational elasticity of network chains, at least at low strains.

This situation is well known for linear polymeric glasses, but for densely crosslinked networks the experimental data are very limited. However, it is not yet clear whether the densely crosslinked polymer networks in the glassy state exhibit, in general, the same mechanical behaviour as linear polymeric glasses. The results given in this Section allow us to answer this question concerning some important mechanical properties.

### 5.2.1 Influence of the Chemical and Topological Structure on the Mechanical Behaviour of Networks

The true stress-strain curves for DGER-DAP ($P = 1$) cured up to $\alpha^{top}$ at different thermal prehistory and temperatures [12, 15] are summarized in Fig. 17 (a, b). These curves are typical, although extension at break $\varepsilon_b$ for the networks cured with mPhDA or DADPhS is usually lower ($\varepsilon_b = 4.5$–$5.5\%$ at $\dot{\varepsilon} = 10^{-2}$ min$^{-1}$ and $T = 25$ °C). The small $\varepsilon_b$ values are not determined by the network structure as was shown in Sect. 5.1.2 and 5.1.3, but they result from the high rate of the fracture processes, which is typical for any glassy substance.

Many characteristic features of the stress-strain curves are rather close to those of linear polymeric glasses. Young moduli E of the considered networks are close to the values known for of low- and high-molecular-weight organic glasses [62].

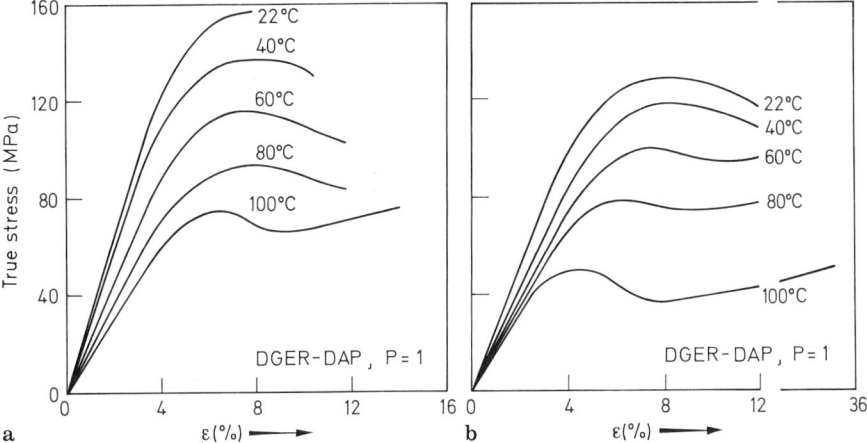

**Fig. 17a and b.** True stress-strain curves (extension rate, $\dot{\varepsilon} = 10^{-2}$ min$^{-1}$) for glassy DGER-DAP (P = 1) network at different test temperatures. annealed (cooling at the rate $\dot{T} = 10$ K/h from 160 °C to 70 °C) and quenched (jump-cooling) samples

The yield stress, $\sigma_y$, is a typical feature of the stress-strain behaviour of networks and only the small values of $\varepsilon_b$ at lower testing temperatures (or high strain rates) do not allow us to reach the macroscopic yield which is demonstrated by the stress-strain curves at higher temperatures (Fig. 17). The values of $\sigma_y$ and $\varepsilon_y$ are very close to those for linear polymer glasses ($\sigma_y^{25} = 9$–15 MPa and $\varepsilon_y^{25} = 4$–6% at $\dot{\varepsilon} = 10^{-1}$–$10^{-3}$ min$^{-1}$). The change of E with temperature corresponds to the thermal expansion coefficient of the polymers and can be described in terms of disharmonic intermolecular vibrations [63]. All these results show that Young moduli of the consideed densely crosslinked networks in the glassy state are determined by Van-der-Waals intermolecular forces and that the deformation of the chemical network does not play any role. Macroscopic yielding in epoxy-amine networks occurs in a very similar manner as in linear polymers. This means that a high concentration of chemical crosslinks does not restrict the yielding at the microscopic level. The difference in the behaviour of networks and linear polymeric glasses is reflected in the absence of the deformation hardening of the former. This feature is probably caused by the difficulties in the reorientation of intercrosslinked chains. However, one can assume that the microscopic yielding is accompanied by relatively large chain elongations and orientation. At present the local deformation processes in glassy networks are less well understood than with linear polymers. However, the measurement of plastic zone sizes (see Sect. 7) of notched epoxy-amine films [64] shows that the maximum local elongation $\varepsilon^*$ of networks is small ($\varepsilon^* \approx 0.3$), which is close to the values of $\varepsilon_b$ for these samples in the rubbery state. The value of $\varepsilon^*$ for glassy polystyrene, measured under the same conditions, is very large ($\varepsilon^* > 1$). These results indicate that even on a local scale of network deformations the orientation of network chains is not as high as in linear polymers.

For the mechanical behaviour of glassy networks, the relations (if it exists) between $\sigma_y$, E and network topological parameters, such as crosslink density, cure conversion, concentration of monocycles and connectivity of chemical crosslinks are important.

Figure 18 shows some mechanical properties of epoxy-amine polymers as a function of initial component ratio P [11, 15]. From Fig. 18, it is clearly seen that neither $\sigma_y$ nor E are really sensitive to the chemical composition of networks within a broad range of P, i.e. of crosslinking density. However, the same networks in the rubbery state are rather sensitive to P and show the highest values of $\sigma_b$ (breaking stress) and $E_\infty$ at P = 1 due to the highest crosslink density. These results again support the idea that the mechanical properties of a glassy network are primarily a function of the packing density of the glassy state. Chemical defects, such as unreacted groups (below a certain extent) do not influence this packing seriously.

Therefore, the increase in the crosslink density itself at the latest stages of the cure process cannot markedly increase the mechanical ($\sigma_y$, E) properties of glassy epoxy-aromatic amine polymers.

The influence of monocycles on network mechanical behaviour is also interesting. The importance of elastically inactive cycles in the behaviour of rubbery networks is well known [13]. Variations of the concentration of monocycles in the considered networks were performed by curing the compounds given in Scheme IV [26].

It was found earlier [65] that the o-diglycidyl ethers exhibited a high probability of cyclization and the cured products contained a certain amount of elastically ineffective loops, but p- and m-diepoxides did not show any tendency to monocyclization.

The shear moduli for different systems measured in the glassy and rubbery state [26] are given in Table 7. It is seen that the shear moduli in the rubbery state are very sensitive to the probability of loop formation (shear moduli change by a factor of 5). However, the glassy shear modulus $G_{25}$ is constant for any measured system. The constancy of the glassy shear modulus corresponds to almost the same packing densities $K_{25}$ (Table 7). The presence of aromatic fragments in the structure linked together by rather flexible aliphatic chains explains the high packing density. Chemical side reactions (aminolysis, etc.), which change the topology of the networks, do not affect the structure of the glassy state. This feature of glassy networks explains the constant level of Young moduli and $\sigma_y$ for glassy networks of different topology.

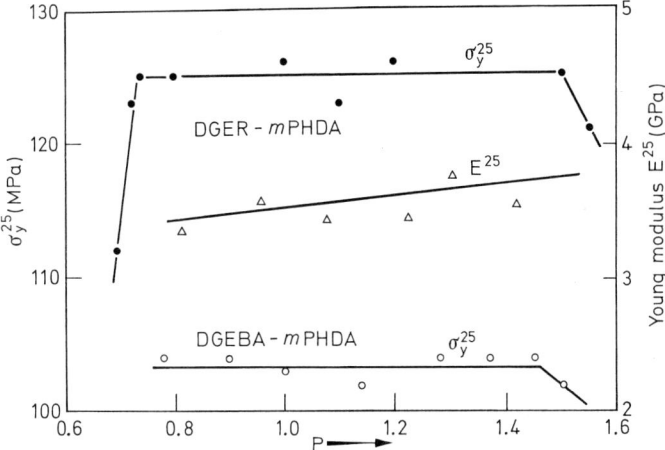

**Fig. 18.** Yield stress $\sigma_y^{25}$ and Young modulus $E^{25}$ for glassy networks. $\dot{\varepsilon} = 10^{-2}$ min$^{-1}$

**Table 7.** Network shear moduli G [26]

| Network | The positions of glycidyl groups in diepoxide | $T_g^\infty$, °C | $G_{T_g^\infty + 40 \,°C}$, MPa | $G_{220\,°C}$, MPa | $G_{25\,°C}$, GPa | Packing Density, $K_{25}$ |
|---|---|---|---|---|---|---|
| DGEHHPhA--mPhDA, P = 1 | ortho | 116 | 6.8 | 8.2 | 1.88 | 0.702 |
| P = 1.3 | ortho | 109 | 4.7 | 5.8 | 1.78 | 0.700 |
| DGEOPhA--mPhDA, P = 1 | ortho | 132 | 5.3 | 5.6 | 1.9 | 0.717 |
| P = 1.3 | ortho | 113 | 2.9 | 2.7 | 1.62 | 0,711 |
| DGEPC-mPhDA, P = 1 | para | 110 | 5.0 | — | 1.60 | 0,724 |
| DGER-mPhDA, P = 1 | metha | 143 | 17.0 | — | 1.8 | 0.719 |
| DGER-DADPhS, P = 1 | metha | 171 | 23.0 | — | 1.5 | 0.717 |
| DGEOPhA--DADPhMOM, P = 1 | ortho | 109 | 11.0 | 11.3 | 1.55 | 0.708 |
| P = 1.3 | ortho | 100 | 4.6 | 5.0 | 1.8 | 0.702 |
| DGER-DADPhS, P = 1.35 | metha | — | 16.0 | — | 1.36 | 0.718 |
| DGEBA-DADPhS P = 1 | para | — | 8.6 | 8.6 | 1.62 | 0.695 |

## 5.3 Post-Cure Thermal Treatment and Mechanical Properties of Networks

Post-cure thermal treatment is known to be a classical way of stabilizing material structure and properties. The processes occurring during treatment may be divided into two categories: physical aging and chemical crosslinking due to side reactions. Physical aging decreases the excess thermodynamic functions (entropy, enthalpy, free volume) of the glass, while the chemical side reactions (see Sect. 2) change the crosslink density of the polymer, which usually makes the packing of the glass denser. The influence of sub-$T_g$ annealing on $C_p$ is shown in Fig. 19.

Structural changes during aging are reflected in the $C_p$ curves as endo-peaks, similarly as for linear polymers [66]. This sub-$T_g$ annealing procedure results in some changes in the density (about 0.1%) and a change in $\sigma_y$ and $E_y$. These changes are seen in Fig. 17. The magnitude of these changes is close to that for linear polymers: $\Delta\sigma_y^{25}$ for networks are equal to 15–20%. It is interesting that fast quench of polymers (see Fig. 17) makes them more plastic.

Both physical aging and chemical side reactions result in a pronounced embrittlement of thermally treated polymers. Young moduli of these polymers usually become a little higher, $\sigma_y$ passes through a maximum (in air much sooner than in vacuum or dry Ar atmosphere) and elongation at break becomes lower. Chemical embrittlement is more pronounced in polymers with an initial excess of the epoxy component.

Post-cure thermal treatment of epoxy-aromatic amine polymers stabilizes mechanical properties to a certain extent. However, the embrittlement makes the treatment harmful from the point of view of engineering material properties.

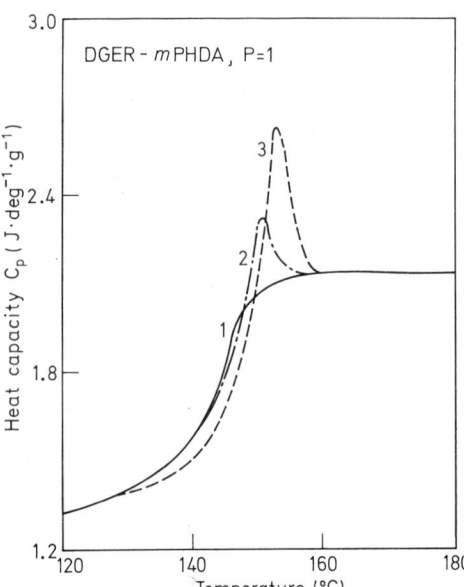

Fig. 19. Change of $C_p$ by sub-$T_g$ annealing. $\dot{T} = 10$ K/min$^{-1}$. 1 Quenched sample; 2 cooled from 110 °C to 25 °C during 9 hours; 3 annealed at T = 120 °C for 90 hours (dry Ar)

## 5.4 Plastic Deformation of Networks

One of the most important features of engineering materials is ductility which depends mainly on the competition between two macroprocesses — plasticity and fracture.

During the last decade, there has been great interest in the plastic deformation of polymeric glasses [67,68,69], but only few papers have dealt with crosslinked systems.

Here, we shall consider several macroscopic features of the plastic deformation of glassy epoxy-aromatic amine networks. Mostly, the tensile or compression deformation has an inhomogeneous character. Usually, diffuse shear zones (or coarse shear bands) are clearly seen at room temperature deformation. Shear zones start from the defects on the sample boundaries or voids (dust) in the bulk. At higher temperatures, the samples are homogeneously deformed with neck formation (DGER-DADPhS, P = 1) [34].

At higher strains, shear zones become broader, become connected with each other and finally cover the whole visible part of the sample, and subsequent deformation proceeds in a "homogeneous manner" on the level of dimensions of optical microscopy. No craze formation has been recorded.

For the networks considered in this paper, the analysis of plastic deformation was carried out [12,70] in terms of Eyring [42], Escaig [68] and Argon [67] models. The changes of activation volume along the stress-strain curve are very similar for all these models (differences are only in the absolute values of $V_a$) and for different network compositions. The activation volume decreases with increasing $\varepsilon$ and reaches a plateau characteristic of the pseudoplastic flow mechanism which is usually observed in linear amorphous polymers [68,71] and in crystals [72]. The increase in $V_a$ with increasing temperature is the same for networks and linear polymeric glasses [71]. It has been found that the plastic deformation of networks correlates well with the Argon model [7,73] and the activation volume $V_a = 450\text{--}500 \times 10^{-30}$ m$^3$ for DGER-$m$PhDA (P = 0.78–1.3). Measurements of activation volume [70] using stress relaxation and the procedure described in Ref. [74] give $V_a = 900 \times 10^{-30}$ m$^3$. The correlation between shear yield stress $\tau_y$ and shear modulus G, extrapolated to T → 0 K in terms of Argon theory, yields $(\tau_y/G)_{T\to 0}^{5/6} = 0.12\text{--}0.13$, which is very close to that of any aromatic polymer [67,73]. The data based on the Argon model are in agreement with the assumption of a rather localized character of plasticity in these network glasses. This localization is even higher ($V_a$ smaller) than for linear polymers like polystyrene or poly(methyl methacrylate) [71,75]. Measurements of other epoxy networks [76] (cured with aliphatic amines) give a very similar picture.

The results again support the point of view that the mechanism of plastic deformation of densely crosslinked networks in the glassy state is rather close to that of linear polymers and that the presence of crosslinks does not basically affect the yield behaviour of the polymer. The localized character of plasticity of networks (small $V_a$) reflects the very local character of nucleation and the growth of plastic nuclei [68,71,75].

The situation resembles the behaviour of molecular motions controlling $T_g$ (see Sect. 4). It was found that the apparent activation energy of the α transition becomes lower with increasing crosslink density (Fig. 13). A very similar situation was detected in the mechanical yielding of epoxides [76], where the activation energy decreased with increasing number of crosslinks per unit volume. The same was found for the epoxies under consideration. The activation enthalpy (Escaig model) and activation energy (Argon model) were the lowest in stoichiometric mixtures [70].

These results make us think that the structural fragments controlling both processes are chemical crosslinks, i.e. amine molecules. Figure 20 shows a possible connection between network topology and yield behaviour. It is clearly seen that the only type of chemical crosslinks whose concentration dependence correlates with the $\sigma_y$ — P curve are the crosslinks participating in three bands N(3). One can assume that the displacement of these crosslinks is a controlling factor in the formation and growth of plastic nuclei, but crosslinks participating in four bands N(4) which determine $T_g$ (see Sect. 4) cannot be displaced under mechanical stress. If that is true, one has a possibility to control various properties of glassy networks simply by changing the ratio of crosslinks of different connectivity. Of course, this assumption should be checked experimentally.

Very interesting features of the plastic deformation of epoxy-aromatic amine networks recently have been found by studying the deformation recovery [77]. Figures 21 and 22 show these results. The sample, deformed in tension or compression, recovers with temperature (temperature scanning regime) in two stages. The first (low temperature) process makes a predominant contribution to the recovery process at low irreversible strains. This contribution is levelled off at strains close to $\varepsilon_y$. At the higher level ($\varepsilon > \varepsilon_y$) of the sample deformation, the main deformation recovery process becomes of the second — high-temperature — type (Fig. 21). These two modes of plastic recovery and, probably, deformation exist only in the glassy state. The loading of the sample at $T \geq T_g$, followed by its cooling down to room temperature, gives only the recovery process of the second type (high temperature). The first mode of plastic deformation recovery is connected with the formation of special shear defects. Figure 22 shows the $C_p$ anomalies — a difference in $C_p$ curves of deformed and undeformed samples — which are a result of sample deformation. A very important feature of these anomalies is their exothermic character in contrast to the endo-type $C_p$ changes at $T_g$. The $C_p$ anomalies were shown [77] to be connected

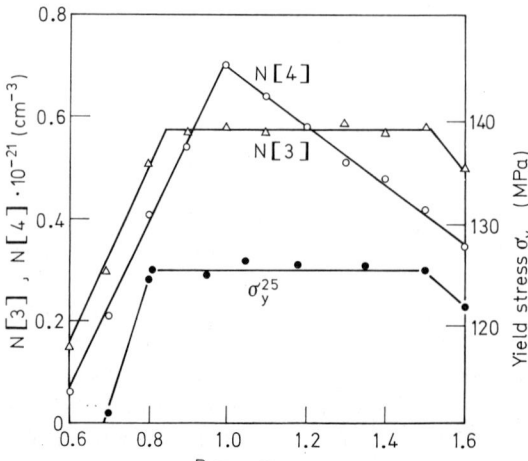

Fig. 20. $\sigma_y^{25}$ and N(4), N(3) knots concentration for completely cured DGER-mPhDA networks

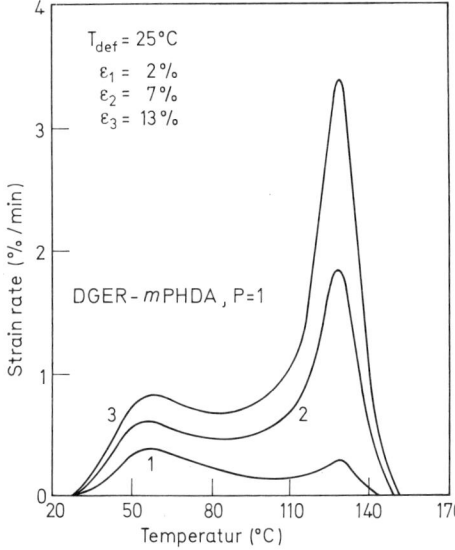

Fig. 21. Size recovery of samples deformed in compression up to different $\varepsilon_{1,2,3}$ (irreversible deformation); $\dot{T} = 5$ deg/min

with the low-temperature deformation recovery process (Fig. 21). It seems possible under deformation of glassy polymers that initially shear defects appear which do not require many conformational rearrangements and which are stabilized by high viscosity of the glass. The number of defects and the extent of their plastic deformation levels off at rather low strains (5–7% at room temperature). The second deformational process involving chain conformational changes starts only after the first process in each locally deformed part of the sample. If this is correct, the first deformation is crystal-like of a shear type, The second process resembles the diffusional plasticity of crystals [78]. The latter corresponds to so-called [79] forced high elasticity of linear glassy polymers. These two modes of plastic deformation of

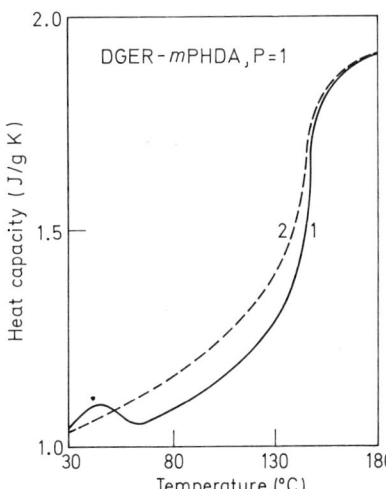

Fig. 22. $C_p$ curves for network samples. (1) deformed at room temperature; (2) undeformed

networks are the same as for polymeric glasses like PS, PC and PETPh [77]. Again, these results show that the plastic deformation of network polymers seems to be very similar to that of any polymeric glass.

# 6 Structure and Properties of Networks Cured at Low $T_{cure}$

The most common cure conditions for the networks under consideration are $T_{cure} < T_g^\infty$ and $T_{cure}$ = const. Under these conditions, the most important feature of the thermosetting process is the vitrification of the reacting mixture which practically quenches chemical crosslinking (see Sect. 2) and changes the physical and mechanical properties of the reacting mixture. In this section, we shall mainly consider the properties of epoxy-aromatic amine glasses prepared at $T_{cure}$ = 40–110 °C.

## 6.1 Quenching of the Cure Process

Vitrification during isothermal cure retards the thermosetting reaction and the conversion $\alpha^{dif}$ becomes lower at lower $T_{cure}$ (Fig. 1). Figure 23 shows the limiting conversions $\alpha^{dif}$ reached by the reacting system DGER-$m$PhDA (P = 1) at different $T_{cure}$. The prolonged exposure of the sample to $T_{cure}$ (several hours) after having reached $\alpha^{dif}$ does not show any detectable change in conversion. For DGER-$m$PDA (P = 1) cure at $T_{cure}$ > 120 °C up to 180 °C for several hours does not show any conversional change. This means that the value $\alpha = 0.92 \pm 0.1$ is the topological limit $\alpha^{top}$ for the curing process of a stoichiometric mixture. This value of $\alpha^{top}$ corresponds well to $\alpha^{top}_{calc}$ = 0.95–0.96, obtained by computer simulation [9]. From Fig. 23, it is clear that if $T_{cure}$ becomes lower than $T_g$ for the initial mixture

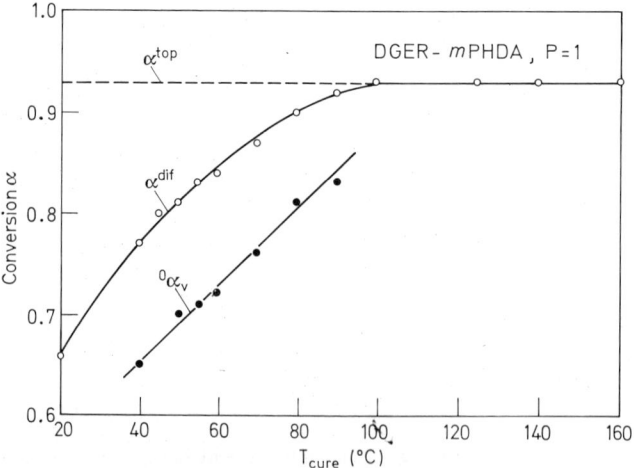

**Fig. 23.** Critical conversions at different $T_{cure}$ (DGER-$m$PDA, P = 1). $°\alpha_v$ — start of vitrification (rise of dynamic Young modulus, Fig. 24)

of reactants, the reaction does not proceed at all, i.e. $\alpha^{dif} = 0$. For DGER-*m*PhDA (P = 1) mixture this value of $T_g$ is $\sim -25\,°C$. This kind of behaviour of thermosetting and polymerizing systems was well represented in the recent series of papers by Gillham et al. [80,81,82] in terms of TTT (Time-Temperature-Transformation) diagrams. The interesting point of Fig. 23 is that $\alpha^{dif}$ becomes equal to $\alpha^{top}$ at $T_{cure} = 110-115\,°C$, which is markedly lower than $T_g$ of the DGER-*m*PhDA (P = 1) network. This means that at $T_{cure} = T_g - (25° \div 30°)$ the reacting groups become mobile and their motion sufficient for curing up to $\alpha^{top}$. Figure 24 shows the mechanical behaviour of the reacting systems. The dynamic Young modulus E' was measured using a mechanical analyzer (DMA-981 DuPont) by placing the reacting mixture inside a small steel spring [83]. This technique allowed us to get E' and E" during the reaction. The picture clearly shows a considerable change of the Young modulus of the reacting samples during vitrification. The quantity $°\tau_v$ represents the time when vitrification just starts and $°\alpha_v$ (conversion corresponding $°\tau_v$) is represented in Fig. 23. No doubt, the value of $°\alpha_v$ depends on the frequency which in our case was $\approx 2-5$ Hz [83]. The vitrification process takes place in the conversion range $\Delta\alpha_v = (\alpha^{dif} - °\alpha_v)$. The time for "crossing" this $\Delta\alpha_v$ interval (Fig. 23) depends on $T_{cure}$.

Figure 24 shows interesting features of vitrification. At low $T_{cure}$, Young moduli of the cured polymers become very high, much higher than the usual values [11,13,15,26,62] of E. The growth of E' during the reaction time proceeds much longer than the growth of $\alpha$ (Fig. 1, 24) at the same $T_{cure}$. Both results are very interesting since they show that the change in the properties of the glassy network strongly depends not only on the chemical composition but also on some structural physical transformations during vitrification. This feature is especially evident at the latest stages of vitrification: $\alpha^{dif}$ becomes lower with decreasing $T_{cure}$ and polymers prepared at low $T_{cure}$ contain a higher concentration of chemical defects — the unreacted groups. E' becomes very high for these chemically defective structures. The comparison of the kinetic curves for $\alpha$ and E' shows that a considerable growth of E' at the given $T_{cure}$

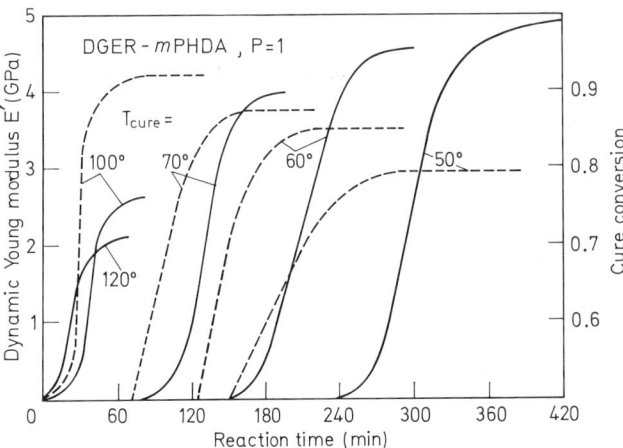

**Fig. 24.** Change of the dynamic Young modulus E' (solid lines) during isothermal cure at different $T_{cure}$. DGER-*m*PhDA (P = 1); dotted lines are the kinetic curves from Fig. 1

continues while the conversion of the reacting system remains practically constant. These results contradict the data given in Fig. 18, where E is practically constant for the cured samples with different P. These results likely show that the structure of the samples prepared at low $T_{cure}$ differs from that formed at high $T_{cure}$. This problem will be discussed below.

## 6.2 Some Properties of Glassy Networks Prepared at Low $T_{cure}$

### 6.2.1 Glass Transition Temperature $T_g^{exp}$

As was mentioned above (Sect. 4), for all considered polymers prepared at $T_{cure} < T_g^\infty$, their experimental glass transition temperature $T_g^{exp}$ is close to their $T_{cure}$. The thermosetting reaction becomes quenched by vitrification, and for a new reinitiation of the cure process the polymer is to be softened by an increase of $T_{cure}$. Experimentally, in all cases, two consecutive processes take place after a sudden increase of $T_{cure}$: (1) the softening of the polymer followed by (2) the next step of cure up to a new $\alpha^{dif}$ (Fig. 25).

The softening of the reacting mixture is clearly demonstrated by a decrease in modulus following the T jump. Only after this softening, the subsequent cure process becomes possible and leads to new values of E' and $\alpha^{dif}$. Figure 26 gives $T_g^{exp}$ values for the polymers cured at different $T_{cure}$; $T_g^{exp}$ follows $T_{cure}$ in a broad range of temperatures. After reaching the condition $T_{cure} \approx T_g^\infty$, the next $T_{cure}$ rise does not change $T_g^{exp}$ [20]. As is shown in Sect. 4, $T_g$ is the highest for stoichiometric mixtures. For nonstoichiometric mixtures, it decreases proportionally to the excess of unreacted groups and to the number of tetra-reacted amine molecules in the polymer. However, the same concentration of unreacted groups results in different $T_g^{exp}$ in polymers prepared at low $T_{cure}$. At the same conversion $\alpha$, the $T_g^{exp}$ of polymers cured at lower $T_{cure}$ is lower than that cured at higher $T_{cure}$. This situation is depicted in Fig. 27 [20, 32]. The solid line denotes $T_g^{exp}$ for polymers prepared at different $T_{cure}$. The samples were cured up to their $\alpha^{dif}$ corresponding to each $T_{cure}$. A dotted line denotes $T_g^{exp}$ for the samples cured at $T_{cure} = 110$ °C. The reaction was stopped by a fast cooling of the reacting mixture. Different conversions were reached at different reaction time.

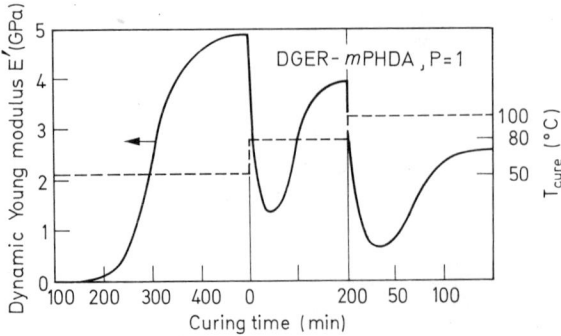

Fig. 25. Change of the dynamic Young modulus with curing time. The sample subsequently cured up to $\alpha^{dif}$. $T_{cure}$ regime is shown by dotted line

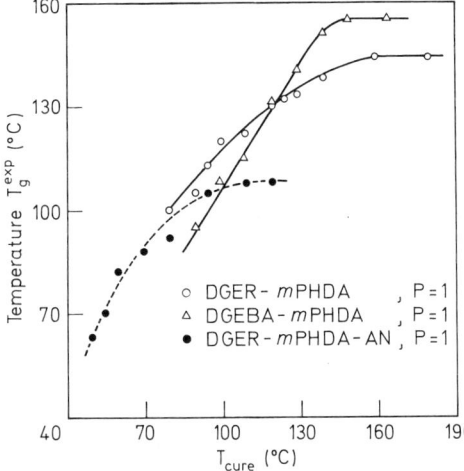

**Fig. 26.** $T_g^{exp}$ as a function of $T_{cure}$. $T_g^{exp}$ measured by DSC; $\dot{T} = 10$ K/min

(Exact values of α after different cure time were determined by measurements of the total heat evolved from the sample during the temperature scan up to 160–180 °C. This value was checked by measurements of α by IR spectroscopy). It is seen that at the same α the values of $T_g^{exp}$ considerably differ which can be explained by structural differences (Fig. 28). The samples cured at low $T_{cure}$ appear as more annealed because they exhibit high endo-peaks on $C_p$ curves. These peaks lower the measured $T_g^{exp}$ values of the samples [44]. However, this structure (with a similar $C_p$ endo-peak) cannot be reached during annealing time close to the cure time at the corresponding $T_{cure}$. The main reason for that is the specificity of glass formation during chemical reaction. The liquid mixture of reactants exists in the liquid state up to the latest stages of the cure process, i.e. up to the conversion $\approx \alpha^{dif}$. In the liquid state, the system is very close to equilibrium due to short relaxation times of liquids and when vitrification sets in the structure of the low-temperature liquid becomes frozen in.

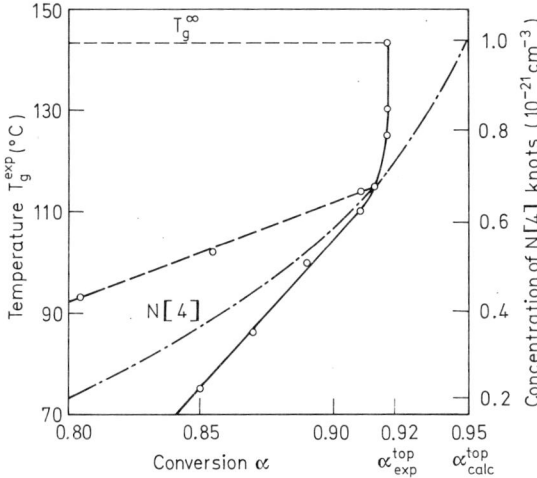

**Fig. 27.** $T_g^{exp}$ and N(4) as a function of conversion α (see text)

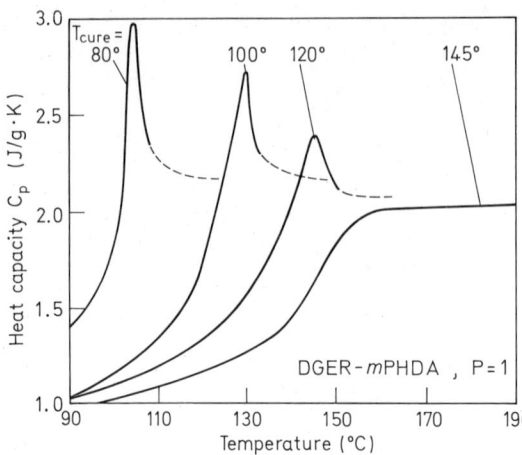

Fig. 28. $C_p$ curves for DGER-$m$PhDA (P = 1) networks prepared at different $T_{cure}$

The situation is different for the glass formed at high $T_{cure}$. Its structure corresponds either to high $T_{cure}$ or to $T_g^\infty$. Annealing of that glass at $T < T_g^{exp}$ usually occurs very slowly [41]. To reach the structure of the epoxy glass prepared at 60 °C, for the glass cured at $T_{cure} = 110$ °C, one should anneal the sample at 60 °C. The relaxation processes in the glassy state at a temperature of 50 K below $T_g^{exp}$ is to proceed for many years. Figure 29 shows the thermodynamic states of the glasses prepared at different $T_{cure}$. Experimentally, the densities and $\beta_g$ of both glasses are the same. This means that the better packing of the glass cured at low $T_{cure}$ is compensated by its lower conversion. Figure 29 clearly shows that in all cases the structure of the glass cured at low $T_{cure}$ is closer to its equilibrium liquid state than that of the glass prepared at high $T_{cure}$. This is why the glass prepared at high $T_{cure}$ cannot be annealed up to the corresponding state of the glass prepared at low $T_{cure}$.

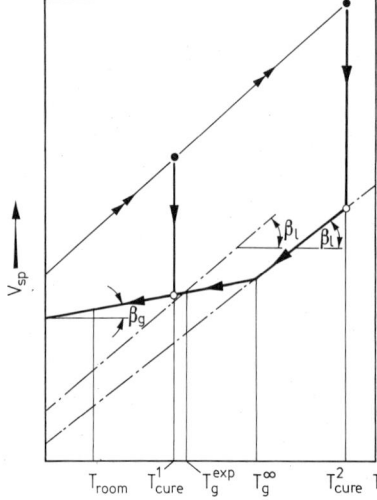

Fig. 29. Change of the specific volume ($V_{sp}$) of a system during cure. $T_{cure}^1 < T_g^\infty < T_{cure}^2$; ○ Vitrification points at $T_{cure}^1$ and $T_{cure}^2$ (conversion $\alpha^{dif}$); ● The start of cure; —·—·— equilibrium liquid state; →→ The heating of initial mixture up to $T_{cure}$, ↓ Cure contraction; ← The cooling of a cured network; $\beta_1$, $\beta_g$ — volume expansion coefficients for liquid (1) and glassy (g) states

Another interesting feature of the $T_g^{exp}$ behaviour is shown in Fig. 27. As was mentioned above (see Fig. 23), the reacting mixture DGER-*m*PDA (P = 1) reaches its $\alpha^{top} = 0.92$–$0.93$ at $T_{cure} \approx 115$ °C. The $T_g^{exp}$ value of the samples is about 120 °C. The next rising of $T_{cure}$ leads to an increase of $T_g^{exp}$ up to $T_g^\infty$, but practically without any chemical reaction.

It seems that some reorganization of the glassy network responsible for $T_g^{exp}$ rise occurs, which is markedly different from a process occurring at lower cure conversions. In Fig. 27 the change in the concentration of crosslinks with a N(4) connectivity with $\alpha$ at the latest stages of cure is shown: the change of N(4) is not as sharp as of the $T_g^{exp}$, although N(4) must play an important role in the $T_g^{exp}$ increase. The $T_g^{exp}$ is close to $T_{cure}$, but these values are not equal (Fig. 26). Up to $T_{cure} \approx 115$ °C, the values of $T_g^{exp}$ are systematically higher than $T_{cure}$. The main reason for $T_g^{exp}$ and $T_{cure}$ inequality is clearly seen from Fig. 30. During the isothermal cure at $T_{cure}^1 < T_g^\infty$, the reacting system becomes glassy and the dynamic Young modulus E' reaches its highest value for $T_{cure}^1$ (right part of Fig. 30). Rather long sample exposure (several tens of hours) at the given $T_{cure}$ after levelling off the modulus does not practically change E'. Subsequent cooling of the sample to room temperature (RT) raises E' due to the sample thermal contraction [63].

For $T_g^{exp}$ measurements, the sample is scanned along the temperature axis (left part of Fig. 30). In this case, the sample retains the value of the Young modulus typical for the glassy state up to $T = T_{cure}^1$. Thereafter the modulus decreases sharply. The difference between $T_{cure}^1$ and $T_g^{exp}$ is clearly seen and is defined by the operational definition of $T_g$. If one defines $T_g$ as the starting point of the E' temperature drop, the relation $T_{cure} = T_g^{exp}$ will hold exactly. From Fig. 30 it is seen that both these temperatures are bound to be connected by the approximate Equation:

$$T_g^{exp} \approx T_{cure} + 0.5\Delta(T_g), \qquad (6)$$

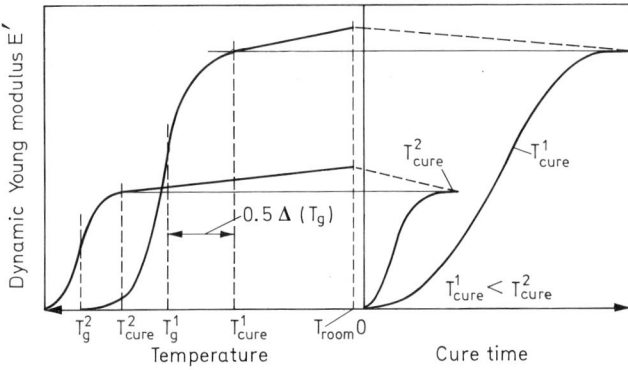

**Fig. 30.** Schematic representation of the change of the dynamic Young modulus E' with time of isothermal cure (right part) and with test temperature (left part). $\Delta(T_g)$ temperature interval of glass transition

where $\Delta(T_g)$ is the temperature interval of the glass transition. Usually, for the considered networks, $\Delta(T_g)$ is 30–60 K, so that:

$$T_g^{exp} = T_{cure} + (15 \div 30 \text{ K}), \tag{7}$$

which is in agreement with experiments [20] (Fig. 26).

### 6.2.2 Mechanical Properties of Samples Prepared at low $T_{cure}$

In this Section are given many examples of unusual mechanical properties of samples prepared at low $T_{cure}$. Figure 24 shows rather high Young moduli of such samples. Some mechanical properties of epoxy-amine networks isothermally cured at different $T_{cure}$ are presented in Fig. 31. At low $T_{cure}$ $E^{25}$ and $\sigma_y^{25}$ values become extremely high and rather extraordinary for any organic glass. The reason is not yet clear. From the data presented in Sect. 5, one can see that the post-cure thermal treatment does not provide samples with such high values of Young moduli compared with those prepared at low $T_{cure}$.

Other interesting examples of the behaviour of the considered samples are given in Fig. 32. Samples with high $T_g$ which are chemically less defective have lower $E^{25}$ values. These anomalies in the mechanical properties of epoxy polymers prepared at low $T_{cure}$ are very difficult to understand in terms of conventional concepts of the relationship between the structure of polymer glasses with their mechanical properties.

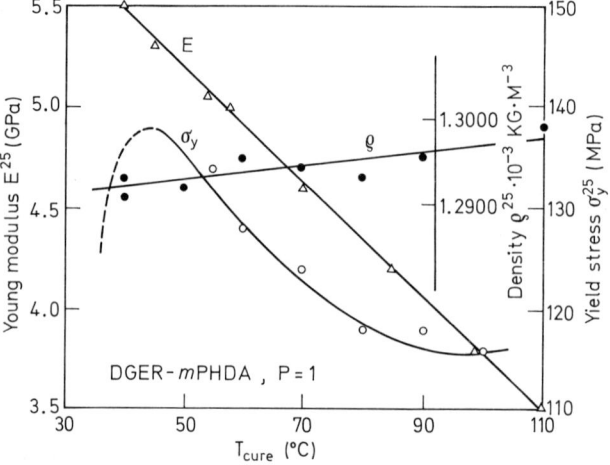

**Fig. 31.** Young modulus $E^{25}$, yield stress $\sigma_y^{25}$ and densities of DGER-$m$PhDA (P = 1) samples prepared at different $T_{cure}$ (cured isothermally up to $\alpha^{dif}$)

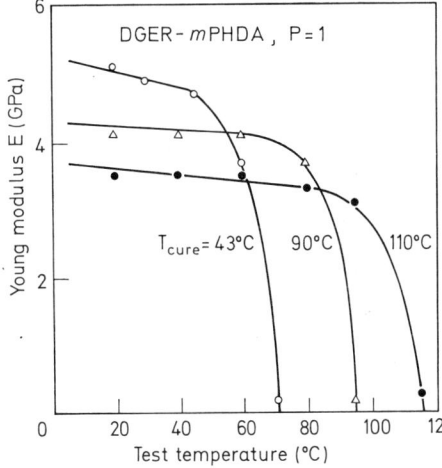

Fig. 32. Young moduli as a function of test temperature; $\dot{\varepsilon} = 10^{-2}$ min$^{-1}$

## 6.3 Structure of Network Glasses Prepared at Low $T_{cure}$

It is clear that the anomalies in the mechanical properties of samples from low $T_{cure}$ compared with samples obtained at high $T_{cure}$ are not due to their chemical differences (concentration of unreacted groups).

Glassy networks prepared at low $T_{cure}$ have some structural specificity as documented by Fig. 33. This figure shows that the shape of the glass transition region is rather different for both types of samples in spite of an almost equal concentration of unreacted groups. Similar structural fragments exist in both types of polymers, but the shape of the α transition for samples from low $T_{cure}$ very different. It is extremely narrow and exhibits a high cooperativity of transition, which is not the case for samples prepared at high $T_{cure}$.

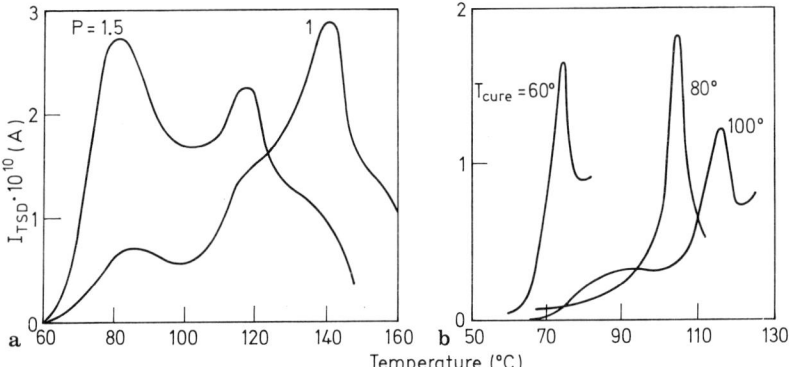

Fig. 33a and b. TSD temperature curves for DGER-$m$PhDA networks prepared under different condition. a) High $T_{cure}$ samples, completely cures up to $\alpha^{top}$ and annealed by slow cooling from 190 °C down to room temperature; b) samples (P = 1) prepared by isothermal cure at different $T_{cure}$ up to $\alpha^{dif}$

It is well known that properties like $T_g$ and the Young modulus are free volume-sensitive [38, 40, 57, 62]. However, the densities of the samples prepared at low $T_{cure}$ are not higher than in the samples prepared at high $T_{cure}$. This makes it difficult to explain the high level of mechanical properties of the glassy samples formed at low $T_{cure}$ in terms of excess free volume.

Thus, the peculiarities of the epoxy glass properties may be due to some specific features of the free volume. For instance, it is possible to explain qualitatively all the unusual properties of epoxy glasses in terms of the size distribution of density fluctuation which may depend on cure conditions [15, 20]. At present, X-ray methods allow to directly measure the spatial distribution of density fluctuations in polymeric glasses [84]. However, it is still unknown how this distribution correlates with properties.

There exists an alternative explanation. Egami, Maeda et al. [85] introduced different kinds of defects of "n", "p" and "τ" type to characterize the glassy state of matter. Defects of n and p-type correspond to negative and positive density fluctuations while the defects of τ-type are the shear defects which do not change the specific volume of the system. Different defects through which the total free volume of a sample is distributed affect different properties. If curing at different $T_{cure}$ leads to polymers with different defect ratios, the differences in the macroscopic behaviour can be explained.

The real structure of glassy networks is still unknown and a number of experiments should be performed to find some new ways of interpreting the results.

## 7 Fracture of Epoxy-Aromatic Amine Networks

### 7.1 Fracture in the Rubbery State

The tension applied to a rubbery network leads to a nonequivalent loading of inter-crosslinked chains due to topological inhomogeneity of the network. This means that in a stressed crosslinked polymer there exists a certain number of overloaded chemical bonds which may behave as active centers of a macroscopic fracture process. The linear relation between macroscopic stress at break and the variance of stress distribution through network chemical bonds [10] obtained from the analysis of the topological diagrams of deformed networks. is in agreement with this suggestion.

Macrofracture is a kinetic process of accumulation of microevents. It can be characterized by some average waiting time or by time-dependent strength of a sample at the given load. If an elementary fracture step is the break of one chemical bond, the average waiting time $\tau_i$ at stress $\sigma_i$ is:

$$\tau_i = \tau_0 \exp\left(\frac{U_0 - \gamma_0 \sigma_i}{RT}\right), \tag{8}$$

where $\tau_0$, $\gamma_0$ are the constants and $U_0$ is the activation energy for breaking a chemical bond.

The Monte-Carlo computer simulation of the macrofracture process shows that each time the process starts from the most overloaded structural elements. The long-term strength of a sample is determined by $\tau_i$ for the first break of a chemical bond, since later the overloading of other bonds increases very quickly. It was found that the kinetic law for the long-term strength of the whole sample was the same as that given by Eq. (8) and $\tau_0$ and $U_0$ of micro- and macroprocesses were also the same. This means that the kinetic parameters of the macrofracture are determined by the initial stress distribution of network chains. This is correct for the rubbery state. The topological structure of a network considerably influences $\gamma$. For networks of high connectivity, the overloading coefficient and $\gamma$ are small.

Theoretical predictions of this behaviour were checked by measurements of the long-term strength of the networks, whose topology was varied by changing the initial ratio $P^{10, 86)}$ of components. It was found that the higher the network connectivity was, the smaller were experimental $\gamma$ values. The lowest $\gamma$ was found in stoichiometric networks. This tendency was in agreement with computer simulation. The experimental activation energy of fracture was the same for all investigated samples and was equal to the activation energy of the chemical cleavage of the weakest chemical bond $C_\beta - C_\alpha(N)$ [17]. These results correspond well to theoretical predictions. However, experimental $\gamma$ values for networks of different P were lower by a factor of 10–25 than the calculated values. This means that the real stress distribution of network chemical chains (bonds) is more inhomogeneous than that obtained by computer simulation. It is likely that some defects and their critical growth play an important role in the fracture process of the considered networks. This problem is still open to further considerations.

## 7.2 Fracture in the Glassy State

Fracture processes become very different in the glassy state as compared with the rubbery state. In the glassy state, the sensitivity of fracture to network topology is lost. The chemical structure of the network, crosslink density and the type of bond overloading do not play a key role and defects of glassy samples become very important.

The analysis of the deformational behaviour of linear polymers allows us to understand the importance of plasticity in fracture behaviour of the glassy state. The same result is valid for the network glasses considered. Fracture processes in glassy polymers of epoxy-aromatic amines type were investigated in Refs. [87 and 64]. The main results of the investigations can be summarized as follows:

— A well-developed plastic zone exists in front of a growing crack tip. The length $L_\perp$ of the plastic zone (along the notch, perpendicular to the tension direction) changes similarly as in glasses of linear polymers. $L_\perp$ in network and linear polymeric glasses is well described by:

$$L_\perp = \frac{\pi K_1^2}{8\sigma_y^2}, \tag{9}$$

where $K_1$ is the stress concentration coefficient and $\sigma_y$ the yield stress.

— The size of the plastic zone parallel to the load direction, $L_\parallel$, shows a different behaviour in the considered networks. The plastic zone is well visible in thin films ($\leq 200$ nm), in contrast to thick samples. In thin films, the size $L_\parallel$ of the zone is equal to the film thickness. This result is in contrast to linear polymeric glasses, where $L_\parallel$ is usually larger than the sample thickness. In networks, $L_\parallel$ does not depend on the length of the notch. However, such a correlation is valid for linear polymers.

In Sect. 5, the macroscopic extension at break of the glassy networks considered was shown to be $\varepsilon_b \approx 4$–$6\%$ and only weakly dependent on the network's chemical composition. However, local plasticity markedly depends on the chemical composition of the network. $L_\parallel$ in the samples with an excess of amine ($P = 1.3$) is about 10 times larger than in the samples with an excess of epoxy groups ($P = 0.8$). This is an additional argument in favour of the assumption of the keyrole of 3-linked chemical crosslinks in network plasticity (see Sect. 5 and Fig. 20). It is evident that the excess of amine in the initial mixture leads to a relatively high concentration of 3-type crosslinks in cured resins.

— The limiting local deformation at break $\varepsilon^*$ is markedly lower ($\varepsilon^* = 0.3$) in networks than in linear polymers (for PS $\varepsilon^* > 1$ [87]).

Thick glassy samples do not show pronounced plasticity and fracture in a more brittle manner. Fracture initiation in glassy networks is in some way connected with the concentration of unreacted groups. Figure 31 shows the growth of $\sigma_y^{25}$ with decreasing $T_{cure}$. When $T_{cure}$ becomes lower than 50 °C, the network samples show macroscopic brittleness at room temperature ($\varepsilon_b < \varepsilon_y$). The reaction conversion in this case is $\alpha^{dif} = 80\%$, which is much higher than the theoretical gelpoint. Thus, the fraction of low-molecular-weight fragments in this sample is rather low, but the concentration of chemical defects (unreacted groups) can be high. These chemical defects are likely to take part in embrittlement of the glassy polymers.

## 8 Conclusions

Studies of epoxy-amine polymers, considered in the present article, offer a rather clear picture of structure-properties relationships for many properties of network polymers depending on their chemical composition in the glassy state near $T_g$ and in the rubbery state. Mechanical properties in the rubbery state at a given chemical composition depend on network topology.

Such properties as $T_g^\infty$, $T_g^{exp}$, $\Delta C_p(T_g)$, $\Delta\beta(T_g)$ are determined by polymer chemical composition and can be calculated using schemes based on property increments.

In contrast to the rubbery state, the properties of glassy networks (mechanical, thermal) at temperatures markedly below $T_g^{exp}$ do not practically depend on chemical composition and cure conversion at the latest stages of the cure process. The sensitivity of the properties to the chemical structure of the network is very weak. The shortness of intercrosslinked chains is displayed only in small values of $\varepsilon_b$, $\varepsilon^*$ and in the absence of deformation hardening for the glassy state of the considered polymers.

The question arises whether all these results are valid for other types of network polymers. It should be noted that the main peculiarity of the structure of the considered networks in the glassy state is the presence of aromatic fragments connected by flexible ether chains. The packing of aromatic fragments determines the network packing density. This is the reason why molecular mobility and related macroproperties are determined by the behaviour of the most immobile network fragments: 4- and 3-linked crosslinks which are the aromatic amine molecules. The kinetic properties are sensitive to the functionality of chemical crosslinks.

The majority of glass properties are determined by the packing density. However, kinetic restrictions for the realization of a dense packing are not operative during network vitrification. Such a glass structure appears to be typical for all kinds of networks composed of flexible and rigid molecular fragments. However, it is not clear whether this approach is applicable to networks consisting of flexible aliphatic chains. The absence of any order or aggregation of polymer fragments in the glassy and rubbery state is another important result. The behaviour of all networks under consideration gives no such result requiring the assumption of a heterogeneous structure in terms of spatial distribution of network chemical crosslinks.

Interesting results have been obtained for glasses cured at low $T_{cure}$. At present it is difficult to state to which extent this behaviour is universal for networks of different chemical structure. One can hope that the low-temperature curing of other systems will also give polymeric glasses with unusual properties.

The network structure does not markedly influence the local sub-$T_g$ molecular motions. This is connected with essential differences in the mobility of aliphatic and aromatic polyfunctional fragments of the network.

However, it becomes clear that the types of motion in the $\alpha$-transition region of crosslinked polymers are more localized and include smaller chain fragments than in linear polymers. This situation is likely to be true for any network, although it depends on the length of intercrosslinked chains and especially on the width of chain length distribution. The same applied to plastic deformation.

Epoxy-aromatic amine networks have turned out to be very convenient objects for obtaining valuable information on the behaviour of crosslinked systems in the glassy and rubbery states. We hope that the approach explained here will yield new results for polymer networks of other chemical structure.

*Acknowledgements*: The author expresses his sincere gratitude to V. M. Asatrian, O. B. Salamatina, A. Ya. Gorenberg for their help in preparing the manuscript and to all his coworkers in the laboratory "Polymer Structure", Institute of Chemical Physics, USSR Academy of Sciences, who participated in the experimental work and discussion of the results.

# 9 References

1. Proceedings of the 1975 International Conference on Composite Materials ICCM/I, (eds.) Scala, E., Anderson, E., Tath, I., Noton, B. R., Vol. 1–2. New York, Metall. Soc. AIME 1976
2. Proceedings of the 1978 International Conference on Composite Materials ICCM/2, (eds.) Noton, B., Signorelly, R., Street, K., Phillips, L., Met. Soc. AIME, New York 1978
3. Advances in Composite Materials. Proceedings of the 1980 International Conference Composite Materials ICCM/3, (eds.) Bunsel, A., Noton, B. R., Paris 1980
4. Progress in Science and Engineering of Composites: (eds.) Hayashi, T., Kawata, K., Umekawa, S., Vol. 1, 2, Japan Soc. Comp. Mater. 1982

5. The Role of Polymeric Matrix in the Processing and Structural Properties of Composite Materials, (eds.) Seferis, J. C., Nicolais, L., Plenum Press, New York 1983
6. Rozenberg, B. A., Oleinik, E. F.: Usp. Khimii, 53, 2, 273, (1984)
7. Dušek, K. in: Rubber-Modified Thermoset Resins, (eds.) Keith Riew, C., Gillham, J. K., Advances in Chemistry Series 208, 3, Am. Chem. Soc., Washington D.C. 1984
8. Knunynts, M. I. et al.: Vysokomol. Soedin., $A25$, 1993 (1983)
9. Topolkaraev, V. A. et al.: Vysokomol. Soedin., $A21$, 1515 (1979); Mekhanika Kompozit. Mater. (SSSR), 2, 195 (1981)
10. Dušek, K., Ilavsky, M.: J. Polym. Sci., Polym. Phys. Ed., 21, 1323 (1983)
11. Artemenko, S. A. et al.: Dokl. Akad. Nauk SSSR, 245, 1139 (1979)
12. Solodysheva, E. S. et al.: Vysokomol. Soedin., $A22$, 1645 (1980)
13. Irzhak, V. I., Rosenberg, B. A., Yenikolopyan, N. S.: Network Polymers (in Russian), Nauka, Moscow (1979)
14. Rozenberg, B. A., Oleinik, E. F., Irzhak, V. I.: Zh. Mendeleev Obsh. SSSR, 23, 3, 272 (1978)
15. Oleinik, E. F.: Pure Appl. Chem., 53, p. 1567 (1981)
16. Pakhomova, L. K. et al.: Vysokomol. Soedin., 820, 7, 554 (1979)
17. Zarkhina, T. S. et al.: Vysokomol. Soedin., $26A$, 811 (1984)
18. Vladimirov, L. V. et al.: Vysokomol. Soedin. $22A1$, p. 225 (1980)
19. Salamatina, O. B. et al.: Vysokomol. Soedin., $25A$, 179 (1983)
20. Oleinik, E. F. et al.: Khimich, Phyzika (SSSR), 3, 885 (1984)
21. Vladimirov, L. V.: Ph. D. thesis, Institute Chem. Phys., USSR Acad. Sci., Moscow (1980)
22. Knunynts, M. I. et al.: Mekh. Kompoz. Mater., 1, 156 (1984)
23. Zhorina, L. A. et al.: Vysokomol. Soedin., $23A$, 2799 (1981)
24. Zarkhina, T. S. et al.: Vysokomol. Soedin., $24A$, 584 (1982)
25. Chepel', L. M. et al.: Vysokomol. Soedin., $25A$, 410 (1983); Vysokomol. Soedin., $26A$, 362 (1984)
26. Ilavsky, M., Bogdanova, L. M., Dušek, K.: J. Polym. Sci. Polym. Phys. Ed., 22, 265 (1984)
27. Horie, K. et al.: J. Pol. Sci., $A-1$, 1357 (1970)
28. Babyevsky, P. G., Gillham, J. K.: J. Appl. Polym. Sci., 17, 2067 (1973)
29. Lunak, S., Vladyka, J., Dušek, K.: Polymer, 19, 931 (1978)
30. Morgan, R. J., O'Neal, J. E. in: Chemistry and Properties of Crosslinked Polymers, Labana, S. S. Editor, 289, Academic Press, New York 1977
31. Topolkaraev, V. A. et al.: Dokl. Akad. Nauk (SSSR) 225, 1124 (1975)
32. Salamatina, O. B. et al.: Vysokomol. Soedin., $23A$, 2360 (1981)
33. Topolkaraev, V. A. et al.: Vysokomol. Soedin., $22A$, 1013 (1980)
34. Topolkaraev, V. A. et al.: Vysokomol. Soedin., $21A$, 299 (1979)
35. Morgan, R. J. in: The Role of the Polymeric Matrix in the Processing and Structural Properties of Composite Materials, (eds.) Seferis, J. C., Nicolais, L., p. 207, Plenum Press, New York 1983
36. Matyi, R. J., Uhlmann, D. R.: J. Polym. Sci., 818, 1053 (1980); Faraday Disc. Soc., 68 (1979)
37. Dušek, K., Plestil, J., Lednicky, F., Lunak, S.: Polymer, 19, 393, (1978)
38. Ferry, J. D.: Viscoelastic Properties of Polymers, New York, 1980
39. Kitaigorodskii, A. I.: Molecular Crystals, Nauka, Moscow 1971
40. Bondi, A.: Physical Properties of Molecular Crystals, Liquids and Glasses, Wiley, New York 1968
41. Kovacs, A. J.: Adv. Polym. Sci., 3, 394 (1963)
42. Brady, T. E., Yeh, G. S. Y.: J. Macromol. Sci. Phys., B, 139, 4, p. 659 (1974)
43. Ponomareva, T. I. et al.: Vysokomol. Soedin., $22A$, 1958 (1980)
44. Moynihan, C. T. et al.: Journ. Amer. Ceram. Soc., 59, 12 (1976)
45. Bulatov, V. V. et al.: Polym. Bull., 13, 21 (1985)
46. Hirai, N., Eyring, H.: J. Polym. Sci. 37, 51 (1959)
47. Wunderlich, B.: J. Phys. Chem., 64, 1052 (1960)
48. Boyer, R. F.: J. Macrom. Sci. — Phys., $B7$, 487 (1973)
49. Vladimirov, L. V., Zelenetskii, A. N., Oleinik, E. F.: Vysokomol. Soedin., $19A$, 2104 (1977)
50. Heijboer, J. in: Physics of Non-Crystalline Solids, (ed.) Prins, J. A., North-Holland, Amsterdam 1965
51. Rudnev, S. N., Oleinik, E. F.: Vysokomol. Soedin., $22A$, 2482 (1980)
52. Chatain, D., Lacabanne, C., Maitrot, M.: Phys. Stat. Stol. (a), 13, 303 (1972)

53. Garroway, A. N., Ritchey, W. M., Moniz, W. B.: Proc. 28th Macromolecular Symposium IUPAC, (eds.) Chien, J. C. W., Lenz, R. W., Vogl, O., p. 1, Univ. Massachus., Amherst 1982
54. Ashok, L. Ch. et al.: Macromolecules, 17, 11, 2399 (1984)
55. Flory, P. J.: J. Macromol. Sci., Phys. *B12* (1976)
56. Fischer, E. W. et al.: Faraday Discuss. Chem. Soc., *68* (1979)
57. Boyer, R. F.: Rubber Chem. Techn.: *36*, 1303 (1963)
58. Boyer, R. F.: Encyclop. Polym. Sci., Techn., *11*, 745 (1977)
59. Robert, G. E., White, E. F. T. in: The Physics of Glassy Polymers, Ed. Haward, R. N., 153, Appl. Sci. Publ., London, 1973
60. Tarasov, V. P. et al.: Vysokomol. Soedin., *24A*, 2379 (1982)
61. Flory, P. J.: Statistical Mechanics of Chain Molecules, Interscience, New York 1969
62. Haward, R. N. in: The Physics of Glassy Polymers, (ed.) Haward, R. N., 3, Appl-Sci. Publ., London 1973
63. Vettegren, V. I., Bronnikov, S.-V., Frenkel, S. Ya.: Vysokomol. Soedin., *26A*, 939 (1981)
64. Pakhomova, L. K. et al.: Vysokomol. Soedin., *23A*, 400 (1981)
65. Chepel', L. M. et al.: Vysokomol. Soedin., *24A*, 8, 1646 (1982)
66. Hedvig, P.: Dielectric Spectroscopy of Polymers, Akad. Kiado, Budapest 1977
67. Argon, A. S. in: Glass Science and Technology, (eds.) Uhlman, D. R., Kreidl, N. J., 79, Acad. Press, New York 1980
68. Escaig, B., Lefebvre, J. M.: Rev. Phys. Appliquee, *13*, 285 (1978)
69. Bowden, P. B. in: The Physics of Glassy Polymers, (ed.) Haward, R. N., 279, Appl. Sci. Publ., London 1973
70. Artemenko, S. A.: Ph. D. Thesis, Inst. Chem. Phys., USSR Acad. Sci., Moscow (1985)
71. Cavrot, J. P. et al.: Mater. Sci. Eng., *36*, 95 (1978)
72. Kocks, U. F., Argon, A. S., Ashby, M. F.: Thermodynamic and Kinetics of Slip, Pergamon Press, New York 1975
73. Argon, A. S., Bessonov, M. I.: Phil. Mag., *35*, 4, 917 (1977)
74. Kubat, J., Righdahl, M.: Intern. J. Polym. Mater., *3*, 287 (1975)
75. Haussy, Y. et al.: J. Polym. Sci., Phys. Edition, *18*, 311 (1980)
76. Yamini, S., Young, R. J.: J. Mat. Sci., *15*, 1814 (1980)
77. Oleinik, E. F. et al.: Dokl. Akad. Nauk SSSR, *286*, 135 (1986)
78. Friedel, J.: Dislocations, Pergamon Press, London (1964)
79. Kargin, V. A., Slonimskii, G. L.: Encyclopedia of Polymer Science and Technology, *8*, 441, New York, J. Wiley 1968
80. Aronhime, M. T., Gillham, J. K.: J. Coat. Techn., *56*, 35 (1984)
81. Enns, J. B., Gillham, J. K.: J. Appl. Polym. Sci., *28*, 2567 (1983)
82. Gillham, J. K. in: Developments in Polymer Characterization, 3, (ed.) Dawkins, J. V., Chap. 5, London, Appl. Sci. Publ. 1982
83. Blaine, R. L., Lofthouse, M. G.: Du Pont Instruments Technical News, Application Brief TA-65
84. Roe, R.-J., Curro, J. J.: Macromolecules, *16*, 428 (1983)
85. Egami, T. in: Molecular and Atomic Glasses, (eds.) O'Reilly, J. M., Goldstein, M., Ann. N.Y. Acad. Sci., *371*, 238 (1981)
86. Knunyants, M. I. et al.: Dokl. Akad. Nauk SSSR, *246* (1979)
87. Manevich, L. I. et al.: Mech. Polym. (USSR), *5*, 860 (1978)
88. Wrasidlo, W.: J. Polym. Sci. A-2, *9*, 1603 (1971)

Editor: K. Dušek
Received September 2, 1985

# Void Growth and Resin Transport During Processing of Thermosetting — Matrix Composites

J. L. Kardos, M. P. Duduković, and R. Dave
Department of Chemical Engineering and Materials Research Laboratory, Washington University, St. Louis, MO 63130, USA

*The fabrication of composite laminates having a thermosetting resin matrix is a complex process. It involves simultaneous heat, mass, and momentum transfer along with chemical reaction in a multiphase system with time-dependent material properties and boundary conditions. Two critical problems, which arise during production of thick structural laminates, are the occurrence of severely detrimental voids and gradients in resin concentration. In order to efficiently manufacture quality parts, on-line control and process optimization are necessary, which in turn require a realistic model of the entire process. In this article we review current progress toward developing accurate void and resin flow portions of this overall process model.*

*Void stability as a function of temperature and pressure is first considered at equilibrium as a bounding behavior for the actual cure cycles. If sufficient moisture is present in the resin, notably high void pressures are possible. Next, the time-dependent stability and growth of voids containing pure water vapor and air/water mixtures is described for a typical commercial curing cycle. The resin pressure early in the cycle and the initial resin moisture content are critical considerations in producing a void-free laminate. A pressure-temperature-humidity stability map is described which identifies conditions for void growth or dissolution throughout the cure cycle.*

*A generalized 3-dimensional resin flow model is summarized, which employs soil mechanics consolidation theory to predict profiles of resin pressure, resin flow velocity, laminate consolidation, and resin content in a curing laminate.*

1 Introduction . . . . . . . . . . . . . . . . . . . . . . . . . . . 102
   1.1 The Autoclave Process . . . . . . . . . . . . . . . . . . . 102
   1.2 Void Evidence . . . . . . . . . . . . . . . . . . . . . . . 104
   1.3 The General Model Framework . . . . . . . . . . . . . . 104

2 Void Formation and Equilibrium Stability . . . . . . . . . . . 105
   2.1 Nucleation of Voids . . . . . . . . . . . . . . . . . . . . . 105
   2.2 Void Stability at Equilibrium . . . . . . . . . . . . . . . . 106

3 Diffusion-Controlled Void Growth . . . . . . . . . . . . . . . 109
   3.1 Problem Definition . . . . . . . . . . . . . . . . . . . . . 109
   3.2 Model Development . . . . . . . . . . . . . . . . . . . . 110
      3.2.1 Assumptions . . . . . . . . . . . . . . . . . . . . . 110
      3.2.2 Governing Transport Equations . . . . . . . . . . 111
      3.2.3 Evaluation of Input Variables . . . . . . . . . . . 112
   3.3 Model Predictions for Void Growth . . . . . . . . . . . . 114

4 Resin and Void Transport . . . . . . . . . . . . . . . . . . . . 119

5 Conclusions . . . . . . . . . . . . . . . . . . . . . . . . . . . 121

6 References . . . . . . . . . . . . . . . . . . . . . . . . . . . . 122

# 1 Introduction

The process of fabricating a high-performance structural composite laminate, such as a 64-ply graphite/epoxy system, currently lacks a firm scientific basis. Although based on sound engineering principles, today's technology is not yielding the part-to-part reliability that will be required of the larger, more complex structures currently on the drawing boards. One of the major problems in achieving this reliability is the occurrence of voids in the final part. Intimately connected with this problem is the mechanism governing resin flow during consolidation of the laminate. Clearly, the void phenomenon occurs during a very complex fabrication process involving heat, mass, and momentum transfer with simultaneous chemical reaction in a multiphase system with time-dependent material properties and boundary conditions. To model such a process by using first principles of transport phenomena is clearly difficult, and solution of the complex differential equations in closed form is not tenable.

In this review we will present an approach to modeling the stability, growth, and transport of voids during autoclave laminate processing. Before presenting the model formulation and some results, it will be helpful to briefly examine the process details and some of the evidence for voids.

## 1.1 The Autoclave Process

One of the most often used production procedures for fabricating a high-performance structural laminate is the Autoclave/Vacuum Degassing (AC/VD) laminating process. In this process, individual prepreg plies are laid up in a prescribed orientation to form a laminate. The laminate is laid against a smooth tool surface and covered with successive layers of glass bleeder fabric, Mylar or Teflon sheets, and finally a vacuum

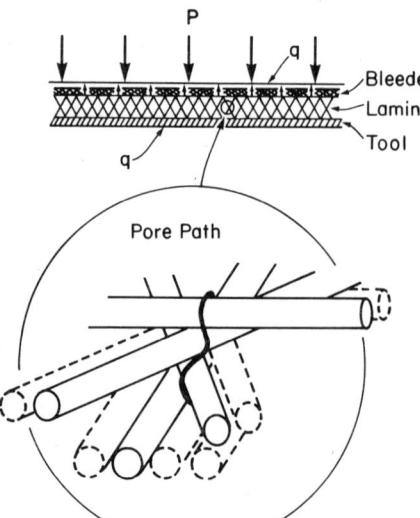

**Fig. 1.** Schematic of laminate lay-up. Insert shows serpentine path which matrix resin and voids might take through connected pores formed by the graphite fibers. Each ply is actually many more fibers thick than is shown

bag. The entire sandwich is shown schematically in Fig. 1 for a [0, ±45, 90]$_{2s}$ laminate. The lower part of the figure depicts a microscopic view of the tortuous pore structure between the fibers for such a laminate; this might be viewed as a repeating unit in the thickness direction for a much thicker laminate of 48 to 64 plies. In reality there are many more fibers in each ply (only one is shown here for simplicity) and so any vertical path is even more tortuous than is depicted here.

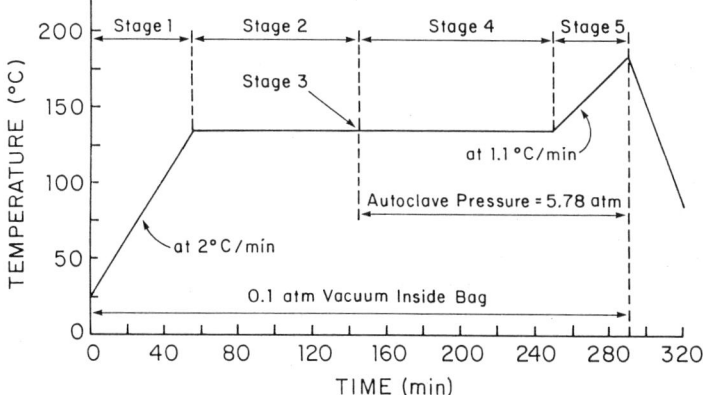

**Fig. 2.** Curing cycle temperature-time profile for typical graphite/epoxy composite in a vacuum bag autoclave process. Autoclave pressure is applied during the 135 °C (275 °F) hold

After bagging, the entire tool is moved into an autoclave, vacuum is applied to the bag, and the temperature is typically increased in the manner shown in Fig. 2 for a T300-5208 system (Thornel 300 fibers/Narmco 5208 epoxy resin system). The 5208 matrix is based on tetraglycidyl methylene dianiline (TGMDA) and diaminodiphenyl sulfone (DDS). What goes on in the laminate during temperature increase and pressure application is a highly complex process. As the temperature is increased, the resin viscosity decreases rapidly and chemical reaction begins. Somewhere between 93 and 135 °C (200 and 275 °F) the resin viscosity reaches a minimum and then begins to increase. However, up to this point, little laminate consolidation has taken place other than that associated with wetting between the plies. During the 135 °C (275 °F) temperature hold, an autoclave pressure of 5.78 atm (85 psi) is applied and laminate consolidation occurs.

The transfer of autoclave pressure to the resin in the laminate does not occur hydrostatically because the resin is not enclosed in a constant-volume system. Flow can occur initially both vertically (thickness direction) and horizontally. Furthermore, the network of fibers can also eventually act as a network of springs to which the vacuum bag and bleeder assembly transfer the load from the autoclave pressure. This load can then be transferred through the fiber network to the tool surface with only a slight frictional loss. As the fiber spring network is loaded and compressed, more resin flow occurs; however at this point, the viscosity is increasing, and the permeability is decreasing, making resin flow more difficult and increasing the pressure levels throughout the laminate. Nonetheless, preliminary results show that initial resin flow is high enough so that only a relatively small fraction of the autoclave pressure is ever generated in the resin [1].

## 1.2 Void Evidence

The occurrence of voids has been thoroughly documented in thick laminates [2]. In almost all cases, they are apparently associated with the prepreg surface. The exact mechanism of void formation is not yet known, but in the most general case it can include mechanical entrapment as well as nucleation of stable voids in the resin phase.

## 1.3 The General Model Framework

The model framework for describing the void problem is schematically shown in Fig. 3. It is, of course, a part of the complete description of the entire processing sequence and, as such, depends on the same material properties and process para-

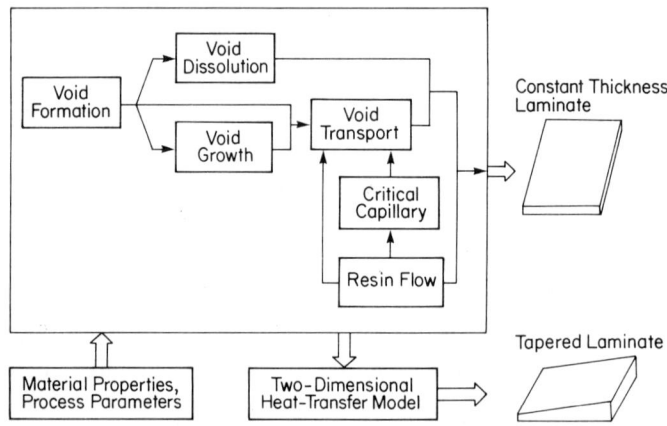

Fig. 3. Schematic of void model framework

meters. Thus, it is intimately tied to both kinetics and viscosity models, of which there are many [3]. It is convenient to consider three phases of the void model; void formation and stability at equilibrium, void growth or dissolution via diffusion, and void transport.

We will present models and results for the first two phases in some detail as well as an initial approach for the resin (and void) transport problem.

## 2 Void Formation and Equilibrium Stability

Voids can be formed either by entrapment of air mechanically or by one of two nucleation processes. Mechanical entrapment could include (1) entrained gas bubbles from the resin mixing operation; (2) bridging voids from large particles or particle clusters (quenched DDS curing agent, airborne particles, or paper release agent); (3) voids from wandering tows, fuzz balls, or broken fibers; (4) air pockets and wrinkles created during lay-up; and (5) ply terminations. Nucleation can occur either homogeneously within the resin or heterogeneously at a resin/fiber or a resin/particle interface. In the case of mechanical entrapment, the initial void distribution is dictated by the processing steps prior to the AC/VD step and must be fed into the model as an experimentally determined parameter.

### 2.1 Nucleation of Voids

Homogeneous nucleation may be described by assuming that critical-size nuclei will be formed from ideal vapor (water or air) at a rate I given by classical nucleation theory [4]. The Equation is

$$I = \left[\frac{P^*}{(2\pi MkT)^{1/2}}\right] 4\pi r^{*2} n \exp\left[\frac{-\Delta F^*}{kT}\right] \quad (1)$$

where
$P^*$ = water vapor pressure, air pressure, or total mixture pressure (water plus air),
$M$ = molecular weight of the vapor phase,
$k$ = Boltzmann constant,
$T$ = absolute temperature,
$r^*$ = radius of the critical nucleus,
$n$ = number of molecules/unit volume in the nucleus, and
$\Delta F^*$ = maximum free energy barrier for the nucleation process.
The free energy barrier $\Delta F^*$ and the critical nucleus size $r^*$ are given by [4]

$$\Delta F^* = \frac{16\pi \gamma_{LV}^3}{3(\Delta F_v)^2} \quad \text{and} \quad r^* = -\frac{2\gamma_{LV}}{\Delta F_v} \quad (2)$$

where $\Delta F_v$ is the free energy change per unit volume for the phase transition (dissolved air or water to air or water vapor), and $\gamma_{LV}$ is the surface energy between the liquid resin and the void nucleus. $\Delta F_v$ can conveniently be approximated by

$$\Delta F_v = \Delta H_v \left(\frac{T_0 - T}{T_0}\right) \quad (3)$$

where $\Delta H_v$ is the heat of transition per unit volume between the two phases, $T_0$ is the equilibrium transition temperature, and $T$ is the actual temperature of the system.

It is far more likely that heterogeneous nucleation plays the governing role in nucleation of voids. The effect of a particle or substrate is to lower the free energy barrier $\Delta F^*$. Thus, Eqs. (1) and (3) remain unchanged and Eq. (2) takes on the new form

$$\Delta F^* = \frac{16\pi\gamma_{sv}^3}{3(\Delta F_v)^2} \frac{(2 + \cos\theta)(1 - \cos\theta)^2}{4} \quad (4)$$

where $\gamma_{sv}$ is the surface energy between the void nucleus and the substrate, and $\theta$ is the contact angle between the void and the substrate.

Equation (1) provides the rate at which stable nuclei will form and demonstrates the clear rate dependence on the temperature, surface energy, and transition enthalpy.

## 2.2 Void Stability at Equilibrium

Whether or not stable nuclei and mechanically trapped voids grow or redissolve depends on several factors. Growth may occur via diffusion of air or water vapor or by agglomeration with neighboring voids. Dissolution may occur if the changes in temperature and pressure cause an increase in solubility in the resin, as we shall see.

Let us first consider the synergistic effect that water has on void stabilization. It is likely that a distribution of air voids occurs at ply interfaces because of pockets, wrinkles, ply ends and particulate bridging. The pressure inside these voids is not sufficient to prevent their collapse upon subsequent pressurization and compaction. However, as water vapor diffuses into the voids or when water vapor voids are nucleated, there will be, at any one temperature, an equilibrium water vapor pressure (and therefore partial pressure in the air-water void) which, under constant total volume conditions, will cause the total pressure in the void to rise above that of a pure air void. When the void pressure equals or exceeds the surrounding resin hydrostatic pressure plus the surface tension forces, the void becomes stable and can even grow. Equation (5) expresses this relationship

$$P_g - P_e = \frac{\gamma_{LV}}{m_{LV}} \quad (5)$$

where $P_g$ and $P_e$ are the void and resin pressures, respectively, $\gamma_{LV}$ is the resin-void surface tension, and $m_{LV}$ is the ratio of void volume to its surface area. Thus the difference in pressure is counterbalanced by the surface tension forces. When the temperature rises for a constant volume system, $P_g$ will rise faster than $P_e$, whereas $\gamma_{LV}$ will decrease slightly. In addition to $P_g$ increasing in accordance with the perfect gas law, the partial pressure of water in the void can rise exponentially because of the temperature effect on the water vapor pressure.

It is instructive to quantitatively examine how $P_g$ might vary as the temperature is increased under equilibrium conditions. Let us assume first that the air-water vapor mixture is an ideal gas and that a Raoult's Law relationship holds for the partial pressure of the water over the water-resin solution at equilibrium. By Raoult's Law

$$p_1 = x_1 P_{0_1} \tag{6}$$

where
$x_1$ = water concentration in the resin-water solution in mole percent,
$P_{0_1}$ = pure component vapor pressure (in this case, water), and
$p_1$ = partial pressure of water in the gas mixture in the void.

The assumption of an ideal solution is obviously not correct, but in the low water concentration range of interest, the resulting error will not be excessively large for a first estimate.

The total pressure in the gas void ($P_G$) is

$$P_G = p_a + p_1 \tag{7}$$

where $p_a$ is the partial pressure of air. At constant temperature,

$$P_G = p_a + x_1 P_{0_1} \tag{8}$$

Assuming that the equilibrium water uptake in the prepreg is 2% by weight results in a mole percent of 0.286 for the TGMDA/DDS resin system. Note that a very small weight percent of the water yields a rather large mole percent. If $p_a$ is initially 101 kPa (14.7 psia) at lay-up and taking $P_{0_1}$ from the steam tables at various temperatures, we can calculate the increase in void pressure as the curing cycle proceeds. The results are shown in Fig. 4 for various equilibrium percentages of water in the

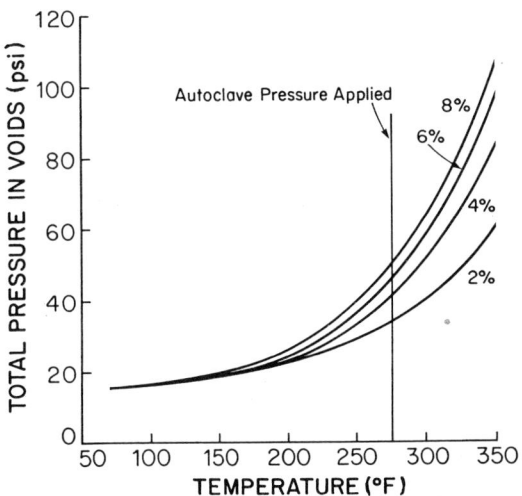

**Fig. 4.** Effect of water on the total void pressure as the temperature increases in a typical curing cycle. Percentages are equilibrium water contents by weight in the resin phase

resin. For a 2% water uptake in the resin, the void pressure can rise to about 228 kPa (33 psia) before autoclave pressure is applied in a typical commercial cycle. Of course, resin viscosity also plays a role in the decision of when to apply the autoclave pressure. Note also that at 177 °C (350 °F), over 414 kPa (60 psia) pressure can be generated in the void.

The calculations shown in Fig. 4 represent an upper bound, but not an unreasonable one. Firstly, although the void was assumed to remain at constant volume, compaction of the laminate is certainly occurring, and unless resin flow into the bleeder is significant, not much resin volume decrease will occur. Furthermore, if volume were not kept constant, then the voids would grow and likely coalesce, a condition of stabilization just as bad, if not worse, than that considered in the calculation. Secondly, it is assumed that Raoult's Law provides the relationship between the water partial pressure in the void and the water dissolved in the resin. It is more likely that Henry's Law, or some other nonideal solution rule, would hold, since the water-resin solution is certainly not an ideal solution; unfortunately there is no way to estimate a priori what nonideal rule should apply. Thus, using Raoult's Law provides an upper bound on the partial pressure, and the calculation represents a worst-case estimate.

It is of interest to examine the effect of void size in conjunction with the void pressure results of Fig. 4. Using Eq. (5) and assuming a spherical shape for the voids, we find

$$P_g = P_e + \frac{6\gamma_{LV}}{d_v} \tag{9}$$

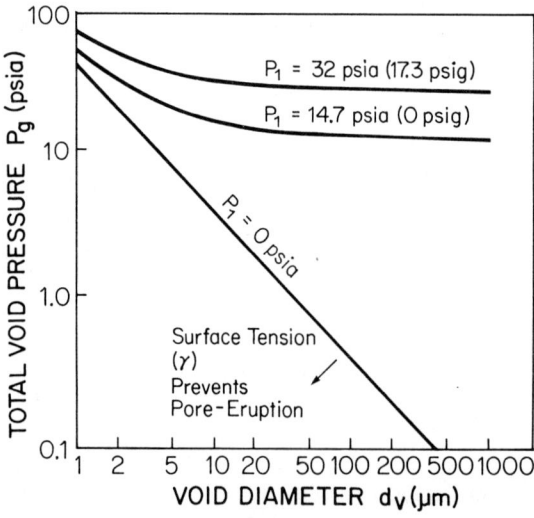

Fig. 5. Void stability map for a typical epoxy resin system. Curves indicate stable void equilibrium states for liquid resin pressures indicated. Growth takes place above the lines and dissolution occurs below the lines for any given resin pressure

where $d_v$ is the void diameter. Using a value for $\gamma_{LV}$ of 50 dynes/cm, one can construct an equilibrium stability map with the resin pressure as a parameter. The results are displayed in Fig. 5 and present some interesting limits to equilibrium void growth. The lines of constant resin pressure represent boundaries above which void growth can occur at that resin pressure, and below which the voids will collapse and eventually dissolve. The region below the line $P_1 = 0$ represents the void pressure necessary to overcome only the resin surface tension forces. For the case of atmospheric pressure in the resin, the total pressure in very small voids must exceed about 276 kPa (40 psia), while larger ones must exceed about 69 kPa (10 psia) in order to grow. As the resin pressure increases, higher total void pressures must be attained in order to create stable voids, particularly for the larger voids.

Preliminary results [1] indicate that resin pressures do not exceed about 103 to 117 kPa (15 to 17 psig), even though autoclave pressures of up to 586 kPa (85 psig) are used. This means that for sufficiently high water contents, void growth will occur and the problem of transporting the voids out of the laminate is extremely important.

# 3 Diffusion-Controlled Void Growth

## 3.1 Problem Definition

In the discussion above, we considered equilibrium void stability. However, actual processing conditions involve changing temperature and pressure with time. While equilibrium calculations provide bounds on void growth, it is the time-dependent growth process that is most important from a process control viewpoint.

One physically realistic possibility involves the growth of a pure water vapor void by diffusion of water from the surrounding liquid resin into the void. Another equally possible mechanism also involves the diffusion process, except that the water enters a void which initially consisted only of entrapped air. In either case it is assumed that water originally dissolved in the resin during prepreg fabrication or storage is the main component contributing to void growth. There are two possibilities for void nucleation. In the case of entrapped air, initial nuclei of finite size always exist. Water dissolved in the resin diffuses into these nuclei whenever conditions are such that this diffusion path is favored. In the absence of air nuclei, water vapor nuclei can form at degrees of supersaturation dictated by Eqs. (1–4). For low molecular weight materials like water, critical-size nuclei contain 50–100 molecules [4] and are so small that their effective diameter is essentially zero.

The exact modeling of void growth would require the solution of many time-dependent coupled partial differential equations. Since many physical properties are not known with sufficient accuracy and since excessive computational time would be required, such an effort does not seem warranted. However, two alternate, simplified approaches seem feasible. One could consider a typical "equivalent" capillary through the plies and examine void growth perpendicular to the plies (see Fig. 1). A knowledge of the pore geometry of the fiber-resin medium and the surface tension, contact angle, viscosity and density of the resin as a function of temperature are required for this approach. A second approach is to consider void growth in an isotropic pseudo-homogeneous medium. This latter approach was selected because experimental evidence indicates that voids do not grow preferentially perpendicular to the plies. This approach then becomes equivalent to bubble growth in a liquid medium.

Numerous papers cover various aspects of bubble growth [5-10]. The classical analysis of Scriven [6] accounts for diffusion, convection transport caused by bubble expansion, and the moving boundary condition at the gas-liquid interface. This approach is superior to the quasistationary equation that considers only diffusion [5] or to the approaches that neglect convective transport but retain boundary movement [11,12]. Subramanian and Weinberg [10] show that the quasistationary approximation [5], which ignores all motion, is more consistent than the approach which retains only the boundary movement. Thus it appears that void growth can be most expeditiously predicted by using the quasistationary assumption [5] along with Scriven's formulation [6]. Neglecting the time derivative in the quasistationary conservation equation leads to the quasi-steady-state approach [7] which was used by Kardos et al. [2] and which leads to the following equation for void growth:

$$\frac{d(d_B^2)}{dt} = 4D\beta \qquad (10)$$

$$t = 0; \quad d_B^2 = 0 \qquad (11)$$

where $d_B$ (cm) is the void diameter, $D$ (cm²/hr) is the water diffusion coefficient and $\beta$ is the growth driving force defined by

$$\beta = \frac{C_\infty - C_{sat}}{\varrho_g} \qquad (12)$$

where $\varrho_g$ (g/cc) is the density of the gas in the void. For constant $D$ and $\beta$, Eq. (10) predicts that the void diameter changes with the square root of time.

$$d_B = 2\sqrt{D\beta t} \qquad (13)$$

## 3.2 Model Development

### 3.2.1 Assumptions

In developing the void growth model, the following simplifying assumptions were made:
1) the void is stagnant between two plies and its center is not moving with respect to fixed coordinates in the laminate.
2) the void is approximately spherical and its effective size is calculated based on an equivalent sphere.
3) There is no interaction between voids (no coalescence).
4) The void is considered to be in an infinite isotropic fluid medium.

5) For pure water voids, void nucleation is instantaneous. For air/water voids, air void nucleation or entrapment has already occurred during layup or at the beginning of the cure cycle.

6) At any given time, the temperature and moisture concentration in the bulk resin are uniform.

7) At each temperature, a pseudo-steady-state is established with respect to the concentration profile; that is, the profile does not change during diffusion at any particular temperature.

8) The viscous, inertia, and surface tension effects are neglected since they are significant only during the initial expansion of the original void nucleus [6]. As the void grows, its growth rapidly becomes limited by the rate of arrival of the diffusing species.

### 3.2.2 Governing Transport Equations

Under the above assumptions, the conservation equation of the diffusing species may be expressed as:

$$\frac{\partial c}{\partial t} + \frac{da}{dt}\frac{a^2}{r^2}\frac{\partial c}{\partial r} = \frac{D}{r^2}\frac{\partial}{\partial r}r^2\frac{\partial c}{\partial r} \tag{14}$$

with the initial and boundary conditions

$$C(0, r) = C_\infty \tag{15a}$$

$$C(t, \infty) = C_\infty \tag{15b}$$

$$C(t, a(t)) = C_{sat} \tag{15c}$$

When the density of the gas changes much less rapidly than the radius of the void, then the equation of continuity which the void radius a(t) satisfies is:

$$\frac{da}{dt} = \frac{1}{\varrho_g} D \frac{\partial c}{\partial r}(t, a(t)) \tag{16}$$

where a (cm) is the void radius and $\varrho_g$ (g/cm³) is the gas density.

The following assymtotic solution for large values of β was developed by Scriven [6]:

$$d_B = 4\beta \sqrt{Dt} \tag{17}$$

The difference in the void diameter predictions between Eqs. (13) and (17) is a factor of $2\sqrt{\beta}$.

The change in diameter per unit time (growth rate) can be obtained by differentiation of Eq. (17), which yields

$$\frac{d(d_B^2)}{dt} = 16D\beta^2 \tag{18}$$

The initial condition is:

$$\text{at } t \equiv 0; \quad d_B^2 = 0 \tag{19}$$

Equations (18) and (19) may now be used to calculate void growth during the cure cycle. The driving force β is adjusted according to the changing temperature and pressure in the laminate, which affect $C_{sat}$ and $\varrho_g$. This is equivalent to assuming that on the time scale of void growth, which is a relatively slow process, all other processes have a chance to reach a new equilibrium. Therefore, pressure and temperature are assumed to be uniform throughout the resin at any moment in time but to vary with time according to the prescribed cure cycle. Thus Eqs. (18) and (19) represent the essence of the model. All that remains is to properly evaluate the model input parameters such as the diffusivity, D, the resin water concentration, $C_\infty$, the interface water concentration, $C_{sat}$, and the gas density, $\varrho_g$, as functions of temperature and pressure during the cure cycle.

### 3.2.3 Evaluation of Input Variables

In the development of the computer code for the void growth model the following input relationships are provided in the program. These could of course be modified to account for different cure cycles or different material systems. The values used in this study are provided as default options in the code.

i) *Vapor Pressure*: The vapor pressure of water within the void is dependent on temperature according to the Clausius-Clapeyron Equation.

$$\frac{dP_{H_2O}}{dT} = \frac{\Delta H_v}{T(v_g - v_L)} \tag{20}$$

Assuming that: i) the heat of vaporization, $\Delta H_v$, is constant, and ii) the vapor phase behaves as an ideal gas, the following dependence of vapor pressure on temperature results:

$$p^*_{H_2O} = \left(p^*_{H_2O_0} \exp \frac{\Delta H_v}{RT_0}\right) \exp \frac{-\Delta H_v}{RT} \tag{21}$$

where $T_0$ (°K) is the boiling point of water at 1 atm (373 °K); $p^*_{H_2O}$ (atm) is the vapor pressure of water; $p^*_{H_2O_0}$ is the water vapor pressure at boiling (1 atm); $\Delta H_v$ (cal/mol) is the heat of vaporization (9720 cal/mol); R (cal/mol °K) is the ideal gas constant (1.987 cal/mol — °K).

Substitution yields

$$p^*_{H_2O} = 4.962 \times 10^5 \exp\left(\frac{-4892}{T}\right) \tag{22}$$

The pressures inside and outside of the void are effectively equal until the resin viscosity becomes so high that viscous effects become important. As the resin proceeds towards solidification, the pressure in the void can rise significantly above the

resin pressure. Surface tension effects are also negligible for voids larger than 100 μm.
ii) *Partial Pressure*: For an air/water void, it is assumed that the void initially contains dry air only. In the case of an air/water mixture in the void, the mole fraction of dry air is

$$y_{air} = \frac{p_0 T}{p T_0} \left(\frac{d_{B_0}}{d_B}\right)^3 \tag{23}$$

where $y_{air}$ is the mole fraction of dry air; p (atm) is the total pressure in the resin; $d_{B_0}$ (cm) is the initial diameter of the void; $p_0$ (atm), $T_0$ (°K) are the initial pressure and temperature in the resin, respectively.

The partial pressure of water within an air/water void is

$$p_{H_2O} = (1 - y_{air}) p \tag{24}$$

where $p_{H_2O}$ (atm) is the partial pressure of water.

For a pure water void, the partial pressure of water equals the total pressure, i.e., $p_{H_2O} = p$.

iii) *Gas Density*: The density of the gas within the void is given by the following equation.

$$\varrho_g = (1 - y_{air}) \frac{M_{H_2O} p}{RT} + y_{air} \frac{M_{air} p}{RT} \tag{25}$$

where $M_{H_2O}$ (gm/mol) and $M_{air}$ (gm/mol) are the molecular weights of water and air, respectively.

For a pure water void, $y_{air} = 0$ in equation (16) results in $\varrho_g = \frac{M_{H_2O} p}{RT}$.

iv) *Solubility*: A simple parabola fits the water solubility data of the resin in the prepreg equilibrated at different relative humidity exposures as follows (13):

$$S_0 = k_1 (RH)_0^2 = k_1 \times 10^4 \left(\frac{p_{H_2O}}{p^*_{H_2O}}\right) \tag{26}$$

where $S_0$ is the solubility (% wt. of water per unit wt. of prepreg), $k_1$ is a constant, and $(RH)_0$ is the relative humidity. For graphite/epoxy prepregs, it is assumed that water is insoluble in the graphite fibers. For the T300/Narmco 5208 system, $k_1 = 5.58 \times 10^{-5}$ [13].

This constant, $k_1$, was determined from data obtained at 25 °C. At higher temperatures, $k_1$ was assumed not to change. This is a useful approximation and the error introduced is small.

v) *Water concentration*: The water concentration in the resin, C (g/cm³), is a function of water solubility in the prepreg, $S_0$ (wt. %), the weight fraction of resin in the prepreg, $W_R$ (g/g), and the resin density, $\varrho_R$ (g/cm³) as follows:

$$C = \frac{S_0}{100} \frac{\varrho_R}{W_R} \tag{27}$$

For the T300/Narmco 5208 system, the water concentrations within the bulk resin and at the void surface were obtained from measured solubility, weight fraction and density data. The water concentration in the bulk resin is dependent only on the initial relative humidity under which the resin was equilibrated and is given by:

$$C_\infty = 2.13 \times 10^{-6} (RH)_0^2 \tag{28}$$

The water concentration at the void surface is a function of both the temperature and partial pressure of water within the void as follows:

$$C_{sat} = 8.651 \times 10^{-14} \exp\left(\frac{9784}{T}\right) p_{H_2O}^2 \tag{29}$$

The interface concentration is called $C_{sat}$ even though an air-water mixture may not be saturated.

vi) *Diffusivity*: The diffusivity of water in the prepreg (either fresh or cured), follows an Arrhenius-type equation [13].

$$D = D_0 \exp\left(-\frac{E_a}{RT}\right) \tag{30}$$

where D (cm²/h) is the diffusivity; $D_0$ (cm²/h) is the pre-exponential constant; $E_a$ (J/mol) is the activation energy for diffusion per mole; R (J/mol °K) is the universal gas constant and T (°K) is the absolute temperature.

For the system T300/Narmco 5208, $D_0 = 0.105$ cm²/h and $(E_a/R) = 2817$ °K for fresh prepreg [13].

vii) *Resin Content*: The fresh prepreg contained 32 wt% resin (the default option in the computer code developed in this study).

viii) *Resin Density*: A resin density, $\varrho_R$, of 1.22 g/cc was used in this study.

ix) *Initial Relative Humidity* $(RH)_0$: This is the humidity at which the prepreg was equilibrated.

If zero initial void diameter is assumed, then no void growth occurs while $C_{sat} > C_\infty$. The time increment from the start of the cure cycle to the moment when $C_{sat} = C_\infty$ is denoted as $t_{BEGIN}$ and is given by:

$$t_{BEGIN} = \frac{T_{CR} - T_{IN}}{R_T \times 60} \tag{31}$$

where $T_{CR}$ (°K) is the temperature at which $C_{sat} = C_\infty$; $T_{IN}$ (°K) is the initial temperature at the start of the cure cycle, and $R_T$ (°K/min) is the heating rate.

## 3.3 Model Predictions for Void Growth

Using the input information described above, void behavior was examined during the various stages of the cycle shown in Fig. 2. Specifically, these stages are as follows:

Stage 1: The temperature is increased from 25 °C (298 °K) to 135 °C (408 °K) at 2 °C/min., yielding a rise time of 55 min. The autoclave pressure is one atm.; the pressure inside the vacuum bag is 0.1 atm. and remains so throughout the entire cycle.

Stage 2: The temperature is held at 135 °C (408 °K) for (a) 15 min., (b) 60 min., and (c) 90 min. The autoclave pressure is still one atm.

Stage 3: The autoclave pressure is increased from one to 5.78 atm. at a constant temperature of 135 °C (408 °K).

Stage 4: The temperature continues to be held constant at 135 °C (408 °K) under a constant autoclave pressure of 5.78 atm. for 105 min.

Stage 5: The temperature is increased from 135 °C (408 °K) to 179 °C (452 °K) at a rate of 1.1 °C/min. under an autoclave pressure of 5.78 atm.; the rise time taken is 40 min.

Because the actual pressure profile in the resin is as yet unknown, it is assumed that the void experiences a resin pressure of 0.1 atm during Stages 1 and 2, which then increases to 5.78 atm during stages 3 to 5. In actuality, the resin never experiences the total autoclave pressure, so 5.78 atm. represents an upper bound.

Figure 6 shows the effect of the processing cycle on void diameter for pure water and air/water voids of 0.1 cm initial diameter under the specified cycle conditions. It was assumed that the air/water void initially consists of pure air, even though there is likely a small but finite water partial pressure. This plot can be divided into the various stages of void growth and dissolution and interpreted as follows. During

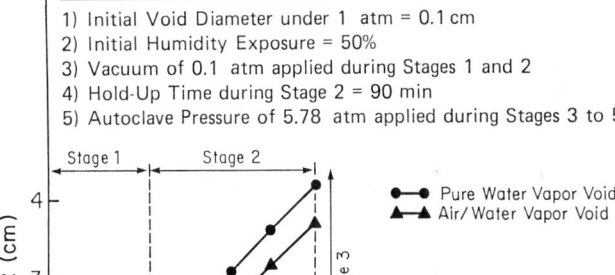

**Fig. 6.** Progression of void growth during the cycle of Figure 2 for both pure water and air/water voids. Note the change when pressure is applied at the end of Stage 2

Stage 1, under constant pressure but increasing temperature, the diffusivity increases exponentially with absolute temperature while $C_{sat}$ decreases exponentially with absolute temperature. However, both these effects favor void growth and a rapid exponential increase in the void diameter results during this stage. During Stage 2, the pressure and temperature are constant and hence the diffusivity, $\varrho_g$, and $C_{sat}$ are constant (for an air/water void, $\varrho_g$ and $C_{sat}$ are not quite constant but the variation is slight). Since $C_\infty$ is also fixed, being dependent only on the initial humidity exposure (Eq. (28)), $\beta$ in Eq. (10) is constant. Since $\beta$ and diffusivity are constant during Stage 2, void growth during this stage is a function of $\sqrt{t}$ only. Thus in Fig. 6, we see a square root relation for void growth with time during Stage 2. During Stage 3, the pressure is increased and the volume decrease is calculated using the ideal gas law. Stage 4 is similar to Stage 2 except that now there is a negative driving force for void growth and hence void dissolution occurs. Similarly, Stage 5 mimics Stage 1, except that void dissolution occurs rather than void growth.

Although the void behavior is calculated through Stage 5, it is likely that the viscosity rises sufficiently high in Stage 4 so that the model assumptions are no longer valid. Thus the void dissolution calculated in Stages 4 and 5 will probably not occur in reality.

An important conclusion from Fig. 6 is that for a small void, it is immaterial wheter or not the initial void contains pure water or an air/water mixture. The diameters at any particular time during the cure cycle are nearly identical when the initial void diameters under 0.1 atm are the same. An air/water void initially containing pure air has a very large driving force for diffusion of water vapor from the resin to the void during the first few minutes of the cycle. This results in diffusion into the void of a large amount of water vapor (relative to the original amount of dry air in the void). Consequently the mole fraction of water vapor in the air/water void quickly approaches unity, and thereafter the rate of diffusion of water vapor across the interface of the air/water void is nearly the same as that for a pure water void.

It is also noteworthy that after the initial period of growth, the higher density of the air/water vapor mixture produces a void which is slightly smaller than the pure water vapor void produced under identical conditions. However, the air/water void cannot completely dissolve during the cure cycle, whereas the water void is capable of complete dissolution.

Figure 7 shows the final diameter of a pure water void at the end of the cure cycle after it has grown from voids of various initial diameters under the conditions specified. For an initial void diameter of zero, the final diameter is about 1.25 cm under the specified conditions of growth. On the other hand, relatively large initial void diameters (0.5 cm) only triple in size.

The final diameter of a pure water void for different initial relative humidities of the resin is shown in Fig. 8. The marked increase in the final void size illustrates the pronounced effect that initial relative humidity exposure has on the final void size. This behavior is described by Equation (28) in which $C_\infty$ (which is fixed during the cure cycle and determines the driving force) increases with the square of the initial relative humidity exposure. Thus, increasing the initial relative humidity by a factor of 2 would result in a 4-fold increase in $C_\infty$. This in turn would increase the driving force 4-fold when the other conditions of growth are kept identical and when $C_{sat} \ll C_\infty$.

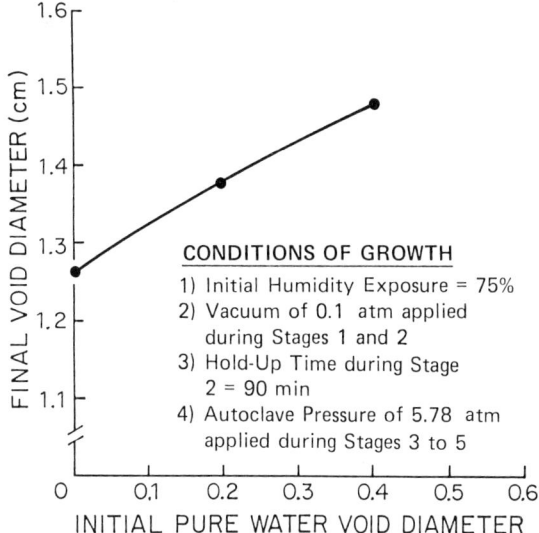

**Fig. 7.** Effect of initial pure water void size on the final void size for the process conditions shown

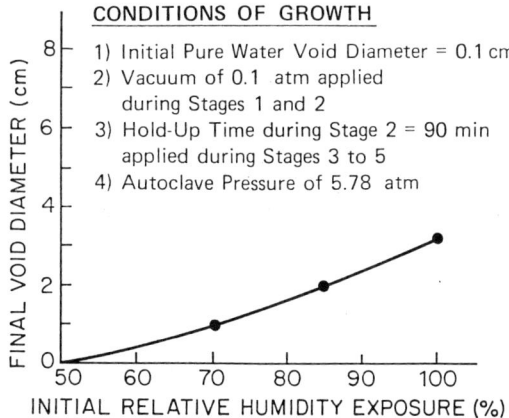

**Fig. 8.** Effect of initial relative humidity exposure of the prepreg on the final void size in the laminate for the process conditions shown

Figure 9 shows the final diameter of the void for different degrees of vacuum applied during Stage 1. It is apparent that the larger the vacuum applied (lower pressure), the greater is the potential for big voids. Referring to equation 29, $C_{sat}$ is directly proportional to the square of the partial pressure of water in the void. The partial pressure of water in the void is equal to the applied pressure for a pure water void, and nearly equal to the applied pressure after a small initial growth period for an air/water void. Thus, reducing the applied pressure by a factor of 2 would result in a 4-fold decrease in $C_{sat}$. This in turn would increase the driving

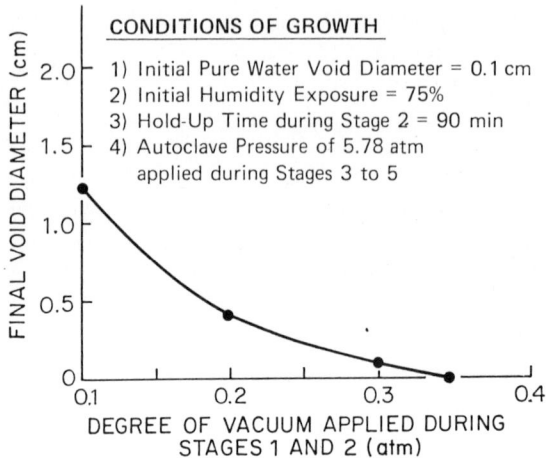

**Fig. 9.** Effect of low matrix pressure (degree of vacuum) on the final void size for the process conditions shown

force. In addition $\varrho_g$ is decreased with decreasing pressure, which further increases $\beta$ and thereby the driving force.

Figures 7–9 identify the two most important parameters that affect the final void diameter, namely (a) the initial relative humidity exposure of the prepreg and (b) the applied pressure during the cure cycle. In terms of industrial processing of laminates, the results of Figure 8 show that the humidity exposure of the resin before and during processing should be kept below about 52% if an initial vacuum of 0.1 atm is to be applied and void-free product is desired. Maintaining this low relative humidity might well be prohibitively expensive, depending upon the climate of the production facility. The alternative to this solution is not to apply an initial vacuum. Figure 9 shows that under the current model assumptions an initial pressure greater than 0.35 atm would prevent voids in the final product.

For a prepreg equilibrated with moisture at a particular relative humidity, in order to prevent the potential for pure water void growth by diffusion at all times and temperatures during the curing cycle, the pressure at all points of the prepreg must satisfy the following inequality:

$$p \geqq 4.962 \times 10^3 \exp\left(\frac{-4892}{T}\right)(RH)_0 \qquad (32)$$

where
$(RH)_0 (\%)$ = initial relative humidity exposure of the prepreg,
$P$ (atm) $= P(t) =$ resin pressure in the prepreg at various times,
$T$ (°K) $= T(t) =$ temperature during the curing cycle.

Equation (32) was derived from the requirement that void growth by diffusion at any temperature cannot occur if the pressure within the void is greater than the saturated vapor pressure at that temperature, i.e., if $C_{sat} \geqq C_\infty$.

While Eq. (32) holds exactly for a pure water void in any system, it would also hold well for small air/water voids after an initial growth period, since it has already been shown in Fig. 6 that the growth pattern of an air/water void is similar to that of a pure water void after an initial growth period.

A plot of Eq. (32) for two relative humidities (50 and 100%) yields a void stability map which is shown in Fig. 10. It is evident from this map that vacuum can be applied without encouraging void growth if such application is coordinated with the temperature of the system. Brown and McKague [14] have experimentally observed that the void content is reduced significantly when pressure is applied to the prepreg early in the cycle, which is in accordance with the stability map.

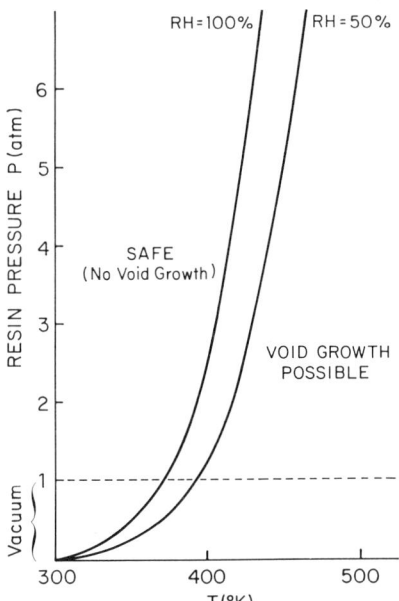

Fig. 10. Void stability map for pure water void formation in epoxy matrices. Note the significant effect of initial relative humidity exposure of the resin

## 4 Resin and Void Transport

For thick epoxy laminates, voids once formed and stabilized can only be removed by resin flow. Furthermore, resin gradients are deleterious to structural laminates. These two key phenomena make an understanding of resin transport vital to the development of any processing model.

In the past, various resin flow models have been proposed (2, 15–19). Two main approaches to predicting resin flow behavior in laminates have been suggested in the literature thus far. In the first case, Kardos et al. [2], Loos and Springer [15], Williams et al. [16] and Gutowski [17] assume that a pressure gradient develops in the laminate both in the vertical and horizontal directions. These approaches describe the resin flow in the laminate in terms of Darcy's Law for flow in porous media, which requires knowledge of the fiber network permeability and resin viscosity. Fiber network permeability is a function of fiber diameter, the porosity or void ratio of the porous medium, and the shape factor of the fibers. Viscosity of the resin is essentially a function of the extent of reaction and temperature. The second major approach is that of Lindt et al. [18] who use lubrication theory approximations to calculate the components of "squeezing" flow created by compaction of the plies. The first approach predicts consolidation of the plies from the top (bleeder surface) down. while the second assumes a plane of symmetry at the horizontal midplane of the laminate. Experimental evidence thus far [19] seems to support the Darcy's Law approach.

Among all the previous models developed so far, only Loos and Springer [15] have considered flow in both the horizontal (laminate plate) and vertical directions. However, in their model the two flows have been decoupled, resulting in different resin pressure profiles along any horizontal line in the laminate for the vertical and the horizontal flows. They have modeled the horizontal flow as laminar flow between ply planes (based on a channel flow equation) with the implicit assumption that a linear pressure gradient exists in the horizontal direction. To apply this equation to the case of flow during consolidation would imply that there is an imaginary source at the vertical midplane of the laminate which is at a higher pressure than the pressure at the edges. Further, in their model, they have considered that the applied pressure on the laminate is borne solely by the pressure in the resin. However, experimental evidence [20] indicates that the resin pressure at the center of the laminate below the ply nearest to the tool surface decreases with time during processing and does not remain constant under constant applied stress.

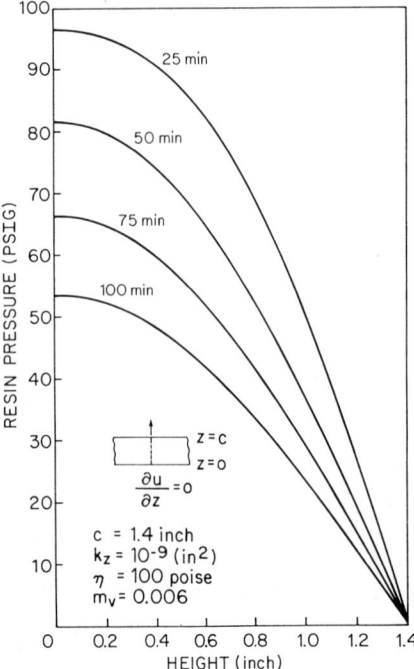

**Fig. 11.** Resin pressure profiles in the laminate thickness direction (vertical) in a 1.4-inch thick unidirectional graphite/epoxy laminate for one-dimensional flow (edge-dammed) under conditions indicated in the Figure

In overcoming the shortcomings of the earlier models, Dave et al. [21] proposed a comprehensive three-dimensional consolidation and resin flow model which can be used to predict the following parameters during cure: (a) the resin pressure and velocity profiles inside the composite as a function of position and time; (b) the consolidation profile of the laminate as a function of position and time; and (c) resin content profile as a function of position and time.

Very briefly, the Dave model considers a force balance on a porous medium (the fiber bed). The total force from the autoclave pressure acting on the medium is countered by both the force due to the spring-like behavior of the fiber network and the hydrostatic force due to the liquid resin pressure within the porous fiber bed. Borrowing from consolidation theories developed for the compaction of soils [22,23], the Dave model describes one-dimensional consolidation with three-dimensional Darcy's Law flow. Numerical solutions were in excellent agreement with closed-form solutions for one- and two-dimensional resin flow cases in which the fiber bed permeabilities and compressibility, as well as the autoclave pressure, are all held constant [21].

Figure 11 depicts the numerical solutions for the time dependence of the resin pressure profile in the vertical direction for one-dimensional flow in the vertical direction (corresponding to an edge-dammed laminate). The laminate is 1.4 inches thick (z direction) and is a unidirectional layup. Values for the specific permeability in the z direction, ($k_z$, in$^2$), the viscosity, ($\eta$, poise), and the coefficient of volume change, $m_v$, are shown in the figure. Clearly, the pressure gradient is not linear as has been assumed in the past.

This model can also provide resin pressure gradients, resin flow rates, consolidation profiles, and, when combined with the void model, void profiles at any point in the laminate. Future papers will deal with these aspects in detail.

# 5 Conclusions

Any on-line process control model used for computer-aided manufacturing of high performance composite laminates must include a thorough treatment of void stability and growth as well as resin transport. These two key components, along with a heat transfer model and additional chemorheological information on kinetics and material properties, should permit optimized production of void-free, controlled thickness parts. Toward this goal a number of advances have been made.

Based strictly on equilibrium considerations, bounds have been set on the stability of voids as a function of temperature and pressure. Although this type of phase map does not depict the time dependency of an actual process, it does provide a limiting scenario toward which the actual process would be heading at any point in the curing cycle. Surprisingly high void pressures are possible if sufficient moisture is present in the resin.

A model and attendent computer code have been constructed to describe time-dependent void growth and stability for any processing cure cycle. Although the analysis is approximate, it does account for the moving void-resin boundary layer and its effect on the concentration profile and diffusion of water in the resin. It was found that for the duration of the cure cycle it makes little difference whether the initial small voids contain pure water or mixtures of air and water.

At the end of a cure cycle, the spread in the final void sizes is not as marked as the spread in the initial void sizes. For any particular cure cycle, the final diameter of the void increases almost linearly but very slowly with increasing initial diameter. Further, even at zero initial void diameter, i.e. void-free resin, the final void diameter may be relatively large due to rapid nucleation and growth. A particularly significant result is that application of vacuum during the initial stage of the cure cycle has the potential of creating large voids. Moreover, the initial moisture content of the prepreg is very important. The void diameter at the end of the cure cycle has a direct second-order dependency on the initial relative humidity exposure of the prepreg.

A pressure-temperature stability map can be constructed as a function of humidity exposure, which identifies the resin pressure values for each temperature below which void growth is possible and above which voids cannot grow but rather tend to collapse.

A generalized three-dimensional resin flow model has been developed, which employs soil mechanics consolidation theory to predict profiles of resin pressure, resin flow velocity, laminate consolidation, and resin content in a curing laminate.

The numerical solutions necessary to solve the practical 3-dimensional problems agree well with the closed-form analytical solutions for simpler one- and two-dimensional cases with constant material properties. The resin pressure gradient in the thickness (vertical) direction for a well-dammed laminate (no horizontal flow) is nonlinear.

*Acknowledgments*: We wish to acknowledge support for this work by the Air Force Wright Aeronautical Laboratories, Materials Laboratory, Wright Patterson AFB, under Contract F33615-80-C-5021 to General Dynamics Corporation (L. McKague-Program Manager) and under Contract F33615-83-C-5088 to McDonnell Aircraft Company, McDonnell Douglas Corporation (F. Campbell-Program Manager). Figures 1, 3, 4, and 5 and some discussion thereof are reproduced under permission from American Society for Testing and Materials, Philadelphia, PA.

# 6 References

1. Hinrichs, R. J.: "Processing Science of Epoxy Resin Composites", Industry Contract Review F33615-80-C-5021, General Dynamics Convair Division, San Diego, CA, Jan. 19, 1982.
2. Kardos, J. L., Duduković, M. P., McKague, E. L. and Lehman, W. M.: Composite Materials: Quality Assurance and Processing, ASTM-STP 797, C. E. Brwoning, Ed., American Society for Testing and Materials, 1983, pp. 96–109.
3. May, C. L., Ed.: Chemorheology of Thermosetting Polymers, ACS Symposium Series, No. 227, American Chemical Society, Washington, DC, 1983.
4. Kingery, W. D.: Introduction to Ceramics, New York: Wiley 1960, Ch. 10, p. 291.
5. Epstein, P. S., and Plesset, M. S.: J. Chem. Phys., *18*, 1505 (1950).
6. Scriven, L. E.: Chem. Eng. Sci., *10*, 1 (1959).
7. Bankoff, S. G.: Advances in Chemical Engineering, 6, New York: Academic Press 1966.
8. Ready, D. E., and Cooper, A. R., Jr.: Chem. Eng. Sci., *21*, 916 (1966).
9. Duda, J. L., and Vrentas, J. S.: AIChE J., *15*, 351 (1969).
10. Subramanian, R. S., and Weinberg, M. C.: J. Chem. Phys., *72*, 6811 (1980).
11. Tao, L. N.: J. Chem. Phys., *69*, 4189 (1978).

12. Tao, L. N.: J. Chem. Phys., *71*, 3455 (1979).
13. Interim Report, "Processing Science of Epoxy Resin Composites", Contract No. F33615-80-C-5021, 9/15/80–10/15/81, Air Force Materials Laboratory, Wright Patterson AFB, OH 45433.
14. Brown, G. G., and McKague, E. L.: "Processing Science of Epoxy Resin Composites", 8th Quarterly Technical Report, Contract No. F33615-80-C-5021, Air Force Materials Laboratory, Wright-Patterson AFB, OH 45433, August 1982.
15. Loos, A. C., and Springer, G. S.: J. Comp. Mats., *17*, 135 (1983).
16. Williams, J., Donnellan, T., and Trabocco, R.: paper presented at 16th National SAMPE Tech. Conf., Albuquerque, Oct. (1984).
17. Gutowski, T. G.: SAMPE Quarterly, July (1985), p. 58.
18. Lindt, J. T.: SAMPE Quarterly, Oct. (1982), p. 14
19. Brand, R. A., and McKague, E. L.: "Processing Science of Epoxy Resin Composites", Tenth Quarterly Report, Contract No. F33615-80-C-5021, Air Force Materials Lab., Wright-Patterson AFB, OH 45433, May (1983).
20. Computer Aided Curing of Composites, McDonnell Douglas Corp., 4th Interim Report, Contract No. F33615-83-C-5088, Air Force Materials Laboratory, Wright-Patterson AFB, OH 45433, April, 1985.
21. Dave, R., Kardos, J. L., and Duduković, M. P.: "A Mathematical Model for Resin Flow During Composite Processing", submitted to Polymer lomp.
22. Terzaghi, K.: Theoretical Soil Mechanics, New York: John Wiley 1943.
23. Taylor, D. W.: Fundamentals of Soil Mechanics, New York: John Wiley 1948.

Prof. K. Dušek (Editor)
Received September 9, 1985

# Physical Aging in Epoxy Matrices and Composites

Eric Siu-Wai Kong
Hewlett-Packard Laboratories, 3500 Deer Creek Road, Palo Alto,
California 94304, USA

*Matrix-dominated physical and mechanical properties of a carbon-fiber-reinforced epoxy composite and a neat epoxy resin have been found to be affected by sub-$T_g$ annealing in an inert dark atmosphere. Postcured specimens of Thornel 300 carbon-fiber/Fiberite 934 epoxy as well as Fiberite 934 epoxy resin were quenched from above $T_g$ and annealed at 140 °C, 110 °C or 80 °C, for times up to $10^5$ minutes. No weight gain or loss was observed during annealing at these temperatures. Significant variations were found in density, modulus, hardness, damping, moisture absorption ability, and thermal expansivity. Moisture/epoxy interactions were also studied. The kinetics of aging as well as the molecular aggregation during this densification process were monitored by differential scanning calorimetry, dynamic mechanical analysis, tensile testing, and solid state nuclear magnetic resonance spectroscopy.*

1 Introduction: Fundamental Aspects of Physical Aging . . . . . . . . . . . 126

2 Experimental . . . . . . . . . . . . . . . . . . . . . . . . . 128

3 Results and Discussion . . . . . . . . . . . . . . . . . . . . 132
   3.1 Stress-Strain Analysis on Neat Resins . . . . . . . . . . . . 132
   3.2 Stress-Strain Analysis on Composites . . . . . . . . . . . . 135
   3.3 Stress Relaxation Analysis on Neat Resins . . . . . . . . . . 138
   3.4 Dynamic Mechanical Analysis on Composites . . . . . . . . . 140
   3.5 Differential Scanning Calorimetry on Neat Resins . . . . . . . 144
   3.6 Thermal Mechanical Analysis on Neat Resins . . . . . . . . . 147
   3.7 Density Measurement on Neat Resins . . . . . . . . . . . . 151
   3.8 Hardness Measurements on the Neat Resins . . . . . . . . . 152
   3.9 Moisture Sorption Kinetics on Neat Resins . . . . . . . . . . 152
   3.10 Proton-decoupled CP/MAS NMR Studies on Neat Resins . . . . 158
   3.11 Solution Carbon-13 NMR on Epoxy Components . . . . . . . 160
   3.12 SAXS Studies on Neat Resins . . . . . . . . . . . . . . . 163

4 Conclusions . . . . . . . . . . . . . . . . . . . . . . . . . 166

5 References . . . . . . . . . . . . . . . . . . . . . . . . . 167

# 1 Introduction: Fundamental Aspects of Physical Aging

In 1931 Simon reported that small molecules in their amorphous solid state are not in thermodynamic equilibrium at temperatures below their glass transition [1]. Such materials are in fact supercooled liquids whose volume, enthalpy, and entropy are greater than they would be in the equilibrium glass. (See Fig. 1).

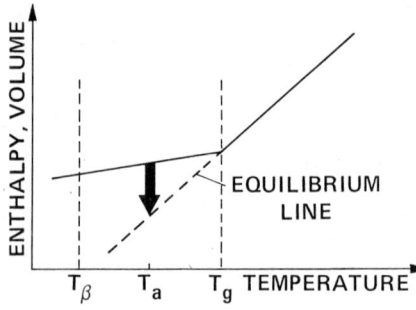

Fig. 1. Origin of physical aging as explained from a thermodynamic point of view

The non-equilibrium glassy state appears to be unstable. Volume relaxation studies of glassy materials have indeed revealed the slow approaches to establish equilibrium, indicating that even below the glass transition temperature, $T_g$, molecular mobility is not quite zero [2-78]. This approach to equilibrium affects many properties of the polymer glass [2-79]. These properties change with time, and the material is said to undergo aging. To distinguish this kind of aging from chemical aging such as thermal degradation or photo-oxidation, Stuart [79] used the term "physical aging" to describe the process.

Physical aging is a continuation of glass formation that sets in around $T_g$. Therefore it affects all those temperature-dependent properties change in the same direction as during cooling through the $T_g$ range. In general, the glassy material becomes stiffer and more brittle; its damping decreases, and so do its creep and stress-relaxation rates [58]. It has also been shown that the dielectric constant and dielectric loss decrease with time during physical aging [80-83].

In this study, results of a systematic investigation of aging phenomenon on one polymer material will be presented. It is a network epoxy which is currently a workhorse resin for aerospace composite materials. In the past, beginning with the work by Kovacs in 1958 [65] until the monograph written by Struik in 1978 [58], emphasis of physical aging studies was on linear macromolecules. It was not until 1978 when the first systematic study on network polymer began [64]. Part of the reasons for this neglect is the difficulty to carefully control the crosslink density of the network and the need for special precautions to avoid chemical aging such as continuous crosslinking reaction in the resin. Hence, researchers tend to avoid such complications which tend to accompany the study of physical aging in thermosets. In this investigation, the epoxy studied was judicially postcured to render a fully-crosslinked network. The aging experiment was designed to be carried out in a dry, dark and inert atmosphere in order to avoid any possible complications that may due to chemical aging.

Physical aging can be explained in a straightforward way from the free-volume concept [84-109]. This is the basic and obvious idea that the chain mobility of macromolecules in a closely packed system is primarily determined by and inversely proportional to the degree of packing of the system. Similarly, mobility can be viewed as directly proportional to free-volume. Throughout this manuscript, the free-volume concept will be used in a qualitative sense. The problem of quantifying free-volume in terms of measurable parameters has not been solved [58]. In fact, efforts in the literature to define free-volume quantitatively have been shown to be self-inconsistent [14]. The quantitative free-volume models will, therefore, not be used in this monograph.

In this investigation, time-dependent variations have been observed in mechanical and physical properties of polymeric network epoxies and also carbon-fiber-reinforced composites with epoxy matrices. These property variations are the result of differences in specimen preparation conditions and/or thermal histories which the materials have experienced [110-117]. In general, with slower cooling rates from above the glass, temperature, and/or increasing the sub-$T_g$ annealing time, the density increases, while impact strength [28], fracture energy [29], ultimate elongation [111], mechanical damping [112], creep rates [58], and stress-relaxation rates [114] decrease.

Of particular concern in the processing and wide-ranging application of structural epoxies is the loss of ductility of such materials on sub-$T_g$ annealing, i.e., thermal aging at temperatures below the glass transition of epoxy resin. This sub-$T_g$ annealing process, more commonly known as "physical aging" [58], is confirmed to be thermoreversible [112]. That is, with a brief anneal at temperatures in excess of the resin $T_g$, the thermal history of an aged epoxy can be erased. A subsequent quench from above $T_g$ would render a "rejuvenated" epoxy. In other words, the polymer embrittles during sub-$T_g$ annealing. However, with an aging history erasure above $T_g$, the ductile behavior can be restored [58].

If epoxies are to be strong candidates as structural matrices for composite materials, it is of primary importance that an understanding of the nature of this volume recovery process as well as an assessment of the magnitude of its effects be achieved. To date, there is a general consensus that the property changes in glassy polymers on sub-$T_g$ annealing are the result of relaxation phenomena associated with the non-equilibrium nature of the glassy state [118-128]. However, a basic understanding of the changes at the molecular level is still lacking. Fortunately, substantial progress has been made in the past few years in characterizing the glassy state from the molecular point of view by powerful techniques such as proton-decoupled cross-polarized magic-angle-spinning (CP/MAS) nuclear magnetic resonance (NMR) spectroscopy [129]. Combining such techniques with other conventional instrumental tools, which measure excess thermodynamic properties, it is now possible to ascertain the changes in properties that can be attributed to the relaxations of excess thermodynamic state functions such as enthalpy and volume. This paper addresses the pertinent relations between excess thermodynamic properties and the time-dependent behavior of epoxy glasses. Also, an attempt is made to describe the molecular nature of this relaxation process.

Moisture is a well-known plasticizer for macromolecules [130-132]. Specifically, water penetrates into an epoxy network and can lower the glass temperature of the resin [131, 133, 134]. In this report, moisture has for the first time been utilized as a probe to characterize densification process during epoxy aging. Also, using the same rationale, heavy water diffused into the epoxy resin is used to analyze the interactions of moisture with the aging polymer by hydrogen-2 (deuterium) NMR spectroscopy.

## 2 Experimental

The epoxy used in this study was Fiberite 934 resin supplied by Fiberite Corporation, Minnesota, U.S.A. The chemical formulation of this resin is shown in Fig. 2. The chemical constituents are 63.2% by weight of tetraglycidyl-4,4'-diaminodiphenyl-methane (TGDDM, tetrafunctional epoxy), 11.2% of diglycidyl orthophthalate (DGOP, difunctional epoxy), 25.3% of the crosslinking agent 4,4'-diaminodiphenyl sulfone (DDS, crosslinker), and 0.4% of the boron trifluoride/ethylamine catalyst complex [135, 136]. (see Fig. 2.)

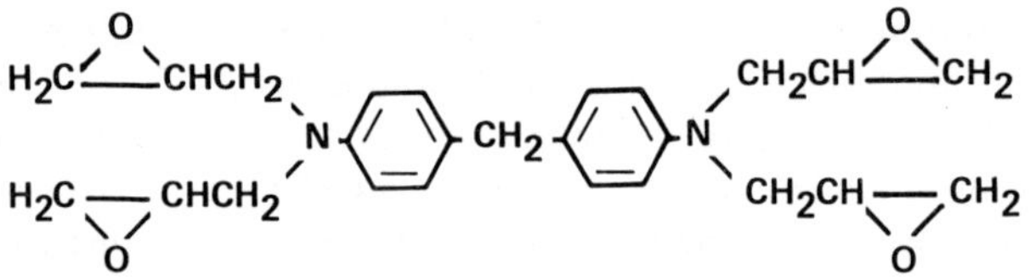

TETRAGLYCIDYL 4,4' DIAMINODIPHENYL METHANE (63.2%)

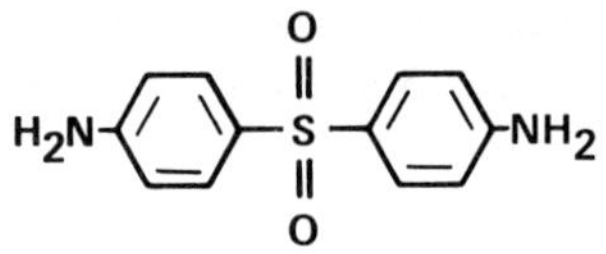

4,4' DIAMINODIPHENYL SULFONE (25.3%)

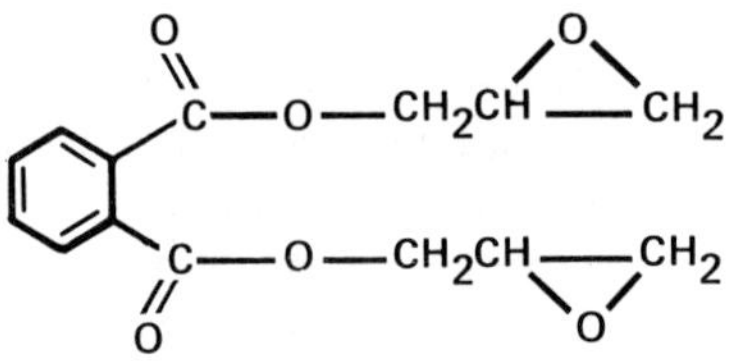

DIGLYCIDYL ORTHOPHTHALATE (11.2%)

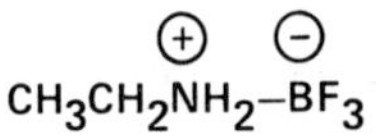

$CH_3CH_2NH_2$–$BF_3$

BORON TRIFLUORIDE/ETHYLAMINE COMPLEX (0.4%)

**Fig. 2.** Chemical constituents of Fiberite 934 epoxy resin

The neat epoxy resin was prepared by casting. The as-received B-stage material was subjected to degasification at 85 °C inside a vacuum oven. The softened resin was then transferred into a preheated silicon-rubber mold. The curing schedule was 121 °C for 2.0 hours, 177 °C for 2.5 hours, followed by a slow cooling at 0.5 °C per minute to room temperature (23.0 °C).

Thornel 300 carbon-fiber-reinforced Fiberite 934 epoxy laminates (ca. 60% fiber and 40% resin by volume) were fabricated from prepreg tapes manufactured by Fiberite Corporation. The details of this fabrication process have been disclosed elsewhere [111]. Panels of suitable lamination sequence were prepared from prepreg types. These panels were cured in an autoclave held for 0.5 hour at 135 °C and 2.0 hour at 180 °C, under 6.8 atm of pressure.

With the exception of five specimens (which were to be tested in the as-fabricated condition), all specimens were postcured for 16 hours at 250 °C, followed by a slow cooling to room temperature at a rate of 0.5 °C per minute. Testing was then performed on the five as-postcured specimens. The other postcured specimens were heated to 260 °C for 20 minutes and then immediately air-quenched to room temperature. Five of these quenched specimens were immediately tested, others were sub-$T_g$ annealed in darkness at either 80, 110 or 140 °C (in nitrogen) for time increments of 10, $10^2$, $10^3$, $10^4$, and up to $10^5$ minutes. The specimens were aged in darkness in order to avoid any chemical aging due to ultraviolet light irradiation. Time zero was taken as the time when a thermocouple placed adjacent to the specimens reached the sub-$T_g$ annealing temperature. At each decade of aging time, five specimens were removed from the environmental chamber and stored at room temperature prior to testing.

In order to demonstrate the "thermoreversibility" aspects of physical aging, the following requenching procedure was carried out. Specifically, some $10^4$ minutes-aged specimens were heated to above $T_g$ for 20 minutes (260 °C), followed by air quenching to room temperature. Five of these requenched specimens were tested immediately, while the rest subjected to "reaging" in darkness at either 80, 110, or 140 °C in nitrogen for time increments of 10, $10^2$, $10^3$, $10^4$, and up to $10^5$ minutes. At least five specimens were tested for each decade of aging time.

Time-dependent stress-strain behavior of the neat resins was studied using an Instron (Model 1122) tensile tester. Dog-boneshaped epoxy specimens were prepared in accordance to ASTM: D1708-66. Strain-rate used was $5 \times 10^{-5}$ s$^{-1}$.

Mechanical properties of the composite materials were tested by a hydraulic-driven MTS tensile tester manufactured by MTS Systems Corporation, Minneapolis, Minnesota. A strain-rate of $5 \times 10^{-5}$ s$^{-1}$ was used. During deformation, the linear actuator position was monitored and controlled by a linear variable differential transformer (LVDT), while strain was measured using MTS-brand axial and diametral strain-gauge extensometers. The axial extensometer serves to measure the tensile deformation in the direction of loading while the diametral extensometer serves to measure the compressive deformation at 90° to the loading axis due to Poisson's contraction. All tensile tests were performed at 23 °C and in accordance to ASTM D3518-76.

Stress relaxation experiments were performed on Instron 1122 tensile tester. Dog-bone-shaped specimens were prepared in accordance to ASTM D1708-66. The specimens were 22.25 mm long (linear section of the dog-bone-shaped specimen),

4.75 mm wide, and 1.50 mm thick. The resin components were prepared as described above and poured into a preheated Dow Corning silicon rubber RTV 3110 mold. In the stress relaxation experiment, the specimens were stretched within a 2-second interval to an elongation of 1.0%. The stress level was then monitored as a function of time under constant elongation. The results are given as the percent of stress relaxation in the first 10 minutes of the experiment. At least five specimens were tested for each decade of sub-$T_g$ annealing time.

Dynamic mechanical analysis was performed on 8-ply Thornel 300/Fiberite 934 composites that were symmetrically reinforced in configuration of $(\pm 45°)_{2s}$. Dynamic mechanical analysis was carried out using a low-frequency, forced-oscillation DMA 981 unit (E.I. du Pont de Nemours and Company) interfaced with a MINC 11 computer (Digital Corporation, Marlboro, Massachusetts). Meaurements were made in nitrogen atmosphere from $-100$ °C to 300 °C at a heating rate of 5 °C per minute. The specimens for the dynamic mechanical experiment were reactangular laminates with a length of 23.8 mm, width 12.7 mm and typical thickness of 1.0 mm.

The Perkin-Elmer DSC-2 differential scanning calorimeter was used to measure the heat capacity of the neat epoxies. The differential scanning calorimetry analysis was performed in nitrogen at a heating rate of 10 °C per minute. The calorimeter was coupled to a "scanning-auto-zero" device which can automate baseline optimization. Each specimen was measured from 50 °C to 280 °C. The specimens were discs of 5 mm in diameter cut from 0.8 mm thick resin-sheet cast using the method described above. The enthalpy relaxation measurements were mady by superimposing the first and second scans for each specimen using a data analysis suggested by Wyzgoski [18,21].

Thermal mechanical analysis was performed using a Perkin-Elmer TMS-2 unit. Thermal expansion behavior was monitored in helium atmosphere from 50 °C to 260 °C at a heating rate of 5 °C per minute. The specimens were discs of 6.0 mm in diameter cut from a 2.5 mm thick resin-plate. Similar to the DSC experiment described earlier, each specimen was scanned twice from 50 °C to 260 °C. After the first scan, a cooling rate of 160 °C per minute was utilized to quench the system from 260 °C to 50 °C. The first and second scans were then superimposed at the high-temperature "rubbery" domain in order to measure the volume recovery during sub-$T_g$ annealing. Thermal expansivity was measured at the linear expansion region below and above the epoxy glass temperature.

Density measurements were made at 23 °C on spherical neat resins of 5 mm diameter using the flotation method in accordance to ASTM: D-1505. The density gradient column (Model DC-1) was supplied by Techne Incorporated, Princeton, New Jersey. Calcium nitrate column was set up which could measure density accurately ranging from 1.210 to 1.290.

Hardness measurements were made at 23 °C on 500 Angstroms gold-decorated epoxy square plates (2.50 cm by 2.50 cm, 2.00 mm thick) using a Leitz miniload micro-hardness tester, supplied by Ernst Leitz Company, Midland, Ontario, Canada. A load of 200 g was applied to the specimen. Experiments were done in accordance to ASTM D-785 and ASTM D-1706 test procedures.

Moisture sorption kinetics by neat epoxies were measured using gravimetric analysis using a Mettler balance which was accurate to $\pm 0.05$ mg. This technique was described in detail elsewhere [137]. Another method was used to monitor the sorption kinetics of heavy water diffusing into neat epoxies. This technique involved the use of solid state hydrogen-2 NMR spectroscopy. By the use of the normalized free induction decay (FID) NMR signal one can readily determine the amount of heavy water sorbed by the epoxy specimen. Cylindrically shaped 20 mm-long epoxy specimens of 5 mm diameter were immersed in heavy water at 23 °C for 60 days and 40 °C for 30 days before the NMR experiment. The hydrogen-2 NMR experiment involved locating the non-spinning heavy-water-saturated solid polymer in a magnetic field of 5 Tesla while pulsing the material with a radio frequency of 30.7 MHz. This technique was used to study moisture-epoxy interactions at the molecular level.

In order to study the molecular aggregation during the volume relaxation of network epoxies, CP/MAS carbon-13 (natural abundance) NMR was utilized. The Hartman-Hahn cross-polarization technique [129] was used with a cross contact time of 1 millisecond for transfer of proton polarization to carbon nuclei. The proton-decoupling was achieved at the radio frequency of 56.4 MHz. Carbon-13 14.2 MHz spectra were measured in a 1.4 Tesla magnetic field. Room temperature (23 °C) experiments were performed at 54.7° MAS at 1 KHz. The spinner was constructed using an Andrew-type rotor driven by compressed air.

A Bruker WM-500 NMR spectrometer was used to study the carbon-13 resonances for the various epoxy components dissolved in deuterated chloroform ($CDCl_3$). Pure components of TGDDM, DDS, and DGOP were dissolved in solvent-containing 10 mm NMR tube. Room temperature carbon-13 NMR spectra were measured at 125 MHz using a superconducting magnetic field of 11.7 Tesla.

The small-angle X-ray scattering (SAXS) patterns were obtained with nickel-filtered copper-K(alpha) radiation at 40 kV and 100 mA using a rotaflex RU-100PL generator supplied by Rigaku-Denki and with a point focusing system so arranged that distances of the first and second pinholes, specimen, and photographic film from the focal spot are 128, 378, 438 and 738 mm respectively. Sizes of the first and second pinholes are 0.5 and 0.2 mm in diameter. The SAXS intensity distributions were detected by a scintillation counter with a pulse-height analyzer. The size of the focal spot was $0.5 \times 5.0$ mm$^2$ on target and $0.07 \times 5.0$ mm$^2$ in projection. The incident X-ray source was collimated by using the following arrangement: the first, second and third slits; the specimen; the counter and scattering slits were placed at 128, 378, 418, 443, 703 and 743 mm from the focal spot respectively. The sizes of the slits of the first and second counters as well as the scattering slit were $0.1 \times 10.0$, $0.1 \times 15.0$, and $0.05 \times 15.0$ mm$^2$, respectively. The intensity distribution was measured by a conventional low-angle X-ray goniometer (Model number 2202 supplied by Rigaku-Denki), using a step-scanning device with a step interval of 0.6 minute, each for a fixed rime of 100 s. The measured scattered intensity distributions were corrected for collimation error by using the weighting function calculated from the Hendricks-Schmidt equations [138] and by using the Schmidt's method of desmearing [139]. Epoxies used in SAXS experiments were cast in class 10 clean room environment.

# 3 Results and Discussion

## 3.1 Stress-Strain Analysis on Neat Resins

Tensile tests were performed on neat epoxy resins in the following conditions: as-cast, as-postcured, as-quenched, and aged at decade increments from 10 to $10^4$ minutes at 140 °C in nitrogen while stored in darkness. A summary of the observed resin stress-strain behavior is shown in Fig. 3. As can be seen, the epoxy polymer was found to be extremely sensitive to its thermal history. The as-cast specimens exhibited the highest value of ultimate tensile strength as well as the greatest values of strain-to-break and toughness. Toughness here is defined as the area under the stress-strain curve, which is different from the dynamic toughness values obtained from impact tests. The postcuring treatment resulted in a significant reduction in these mechanical properties. This effect is undoubtedly due to the crosslinking reactions in the thermoset.

Oddly enough, the postcured specimens given an air-quench from above $T_g$ exhibited a loss in strength, ductility and toughness significantly greater than that of the as-postcured specimens (Fig. 3). This observation was unexpected, based on the free-volume concept. A rapid quench will result in a larger deviation from the equilibrium glassy state; thus, a relatively large amount of free volume will be frozen into the epoxy. Because more free volume can be interpreted to mean higher chain mobility and shorter molecular relaxation time, an increase in free volume was anticipated to result in an increase in epoxy tensile properties instead of a severe decrease.

Quenched specimens given a brief thermal annealing at 140 °C for 10 minutes were found to exhibit toughness similar to that observed for as-postcured specimens (Table 1). Even though the strength for 10 minutes annealed specimens was not

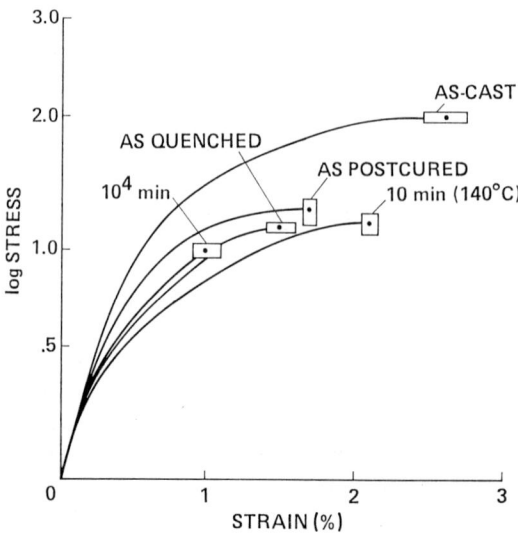

**Fig. 3.** Stress-strain behavior of Fiberite 934 resin as a function of thermal history

## Table 1.

| THERMAL HISTORY | AS-CAST | POSTCURING 16 hr 523K (250°C) | ANNEALING 533K (20 min) +QUENCHING 296K (23°C) | SUB-Tg ANNEALING ||||
|---|---|---|---|---|---|---|---|
| | | | | 10 min 413K (140°C) $N_2$ atm | $10^2$ min 413K (140°C) $N_2$ atm | $10^3$ min 413K (140°C) $N_2$ atm | $10^4$ min 413K (140°C) $N_2$ atm |
| UTS, MPa | 102.20 ± 1.16 | 17.71 ± 1.14 | 13.76 ± 1.13 | 14.73 ± 1.11 | 13.30 ± 1.11 | 11.66 ± 1.19 | 9.50 ± 1.01 |
| $\epsilon_B$, % | 2.6 ± 0.82 | 1.7 ± 0.90 | 1.5 ± 0.90 | 2.1 ± 0.58 | 1.6 ± 0.55 | 1.2 ± 0.55 | 1.0 ± 0.50 |
| TOUGHNESS $J/cm^3$ | 2.69 | 0.30 | 0.21 | 0.31 | 0.10 | 0.07 | 0.06 |
| E, MPa | 13000 | 12381 | 11063 | 11817 | 8965 | 8666 | 10399 |
| $\sigma_y$, MPa | 7.59 | 4.47 | 3.55 | 2.66 | 2.37 | 2.00 | 1.80 |

totally restored compared to the as-postcured specimens, the ductility was much improved. One explanation for these observations is the presence of residual thermal stresses, which can develop in the bulk material as the result of rapid thermal changes. The skin and the core of the bulk epoxy would experience different cooling rates during the air-quench, and rapid cooling of the specimen would not permit the time-dependent relaxation of these stresses. The fact that a brief thermal annealing results in restoration of epoxy tensile properties suggests that the residual thermal stresses have been removed and have not caused irreversible damage in the specimen.

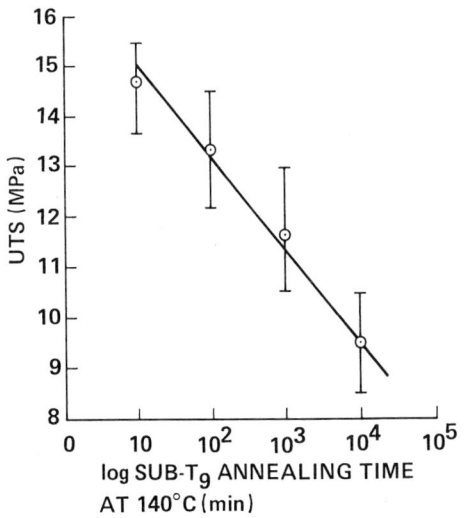

Fig. 4. Ultimate tensile strength of fully crosslinked epoxy as a function of aging time

Thermal annealing at 140 °C in an inert dark atmosphere resulted in decreases in strength, ductility, and toughness, as seen in Fig. 3 and Table 1. These changes are attributed to physical aging processes occuring in the glassy polymer. As an additional check to assure that compositional changes were not occuring in the polymer with thermal exposure, specimen weights were followed. No resolvable weight change was observed in any of the aged specimens.

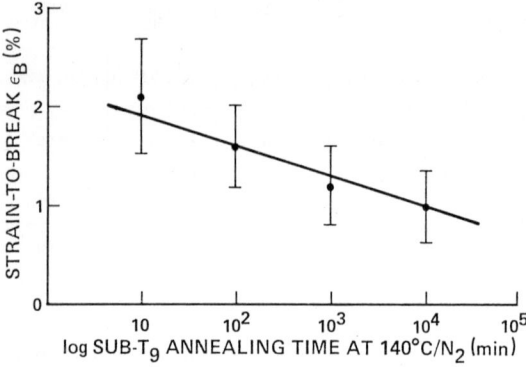

**Fig. 5.** Ductility of Fiberite 934 epoxy as a function of aging time

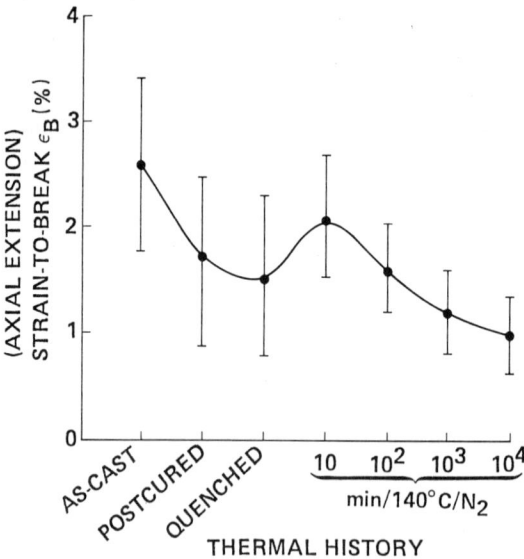

**Fig. 6.** Ductility of Fiberite 934 epoxy as a function of thermal history

The effects of physical aging at 140 °C on ultimate tensile strength and strain-to-break of Fiberite 934 epoxies are shown in Figs. 4 and 5, respectively. The effects of thermal history on the ductility of network epoxies are summarized in Fig. 6. The decreases in ultimate tensile strength and strain-to-break appear to be linear as a function of logarithmic aging time. Toughness and yield strength also decreased with time. The modulus varied somewhat erratically, but was roughly constant. (see Table 1.)

## 3.2 Stress-Strain Analysis on Composites

The composite with epoxy-matrix exhibits mechanical properties which are sensitive to its thermal history. Figure 7 shows that the as-received specimens exhibit by far the greatest degree of toughness. For the as-received materials, ultimate tensile strength averaged 162.60 ± 0.76 MPa (95% confidence intervals) whereas strain-to-break averaged 2.44 ± 0.74%.

Thermal analysis by differential scanning calorimetry on as-received materials indicated that the matrix was not fully cured: an exotherm was detected the first scan from room temperature to 300 °C. A 250 °C/16 hours was therefore necessary for the specimens in order to complete all the crosslinking reactions. After the postcuring, differential scanning calorimetry confirmed a fully cured system which exhibit a regular step-function increase in heat capacity at $T_g$. The calorimetry results are shown in Fig. 8.

Postcuring embrittled the epoxy-matrix composite (Fig. 7). After postcuring the material, ultimate tensile strength decreased to 148.60 ± 6.41 MPa (Fig. 7) and strain-to-break decreased to 1.04 ± 0.07% (Fig. 7). This postcuring schedule was necessary because we wanted to eliminate the possibility of continuous crosslinking reactions (chemical aging) during our subsequent sub-$T_g$ annealing (physical aging) experiments. It has been demonstrated that chemical aging processes such as continuation of crosslinking reactions could contribute to the embrittlement of polymeric resins [140].

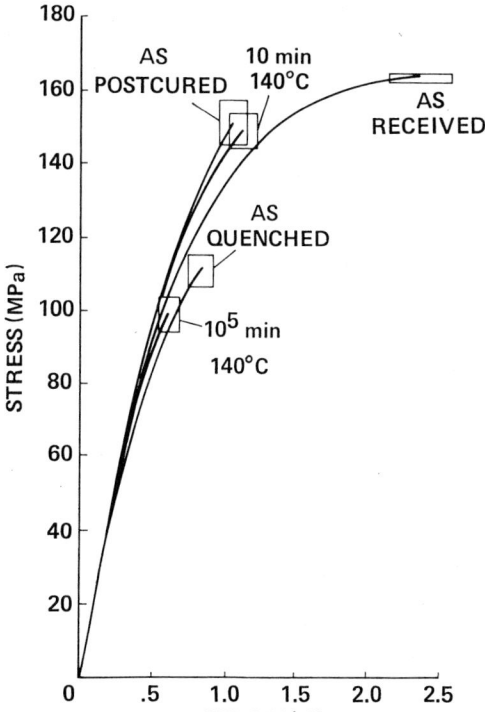

**Fig. 7.** Comparison of stress-strain curves for carbon-fiber-reinforced epoxy composites of different thermal histories. Error rectangles were drawn to indicate a 95% confidence level for both stress and strain

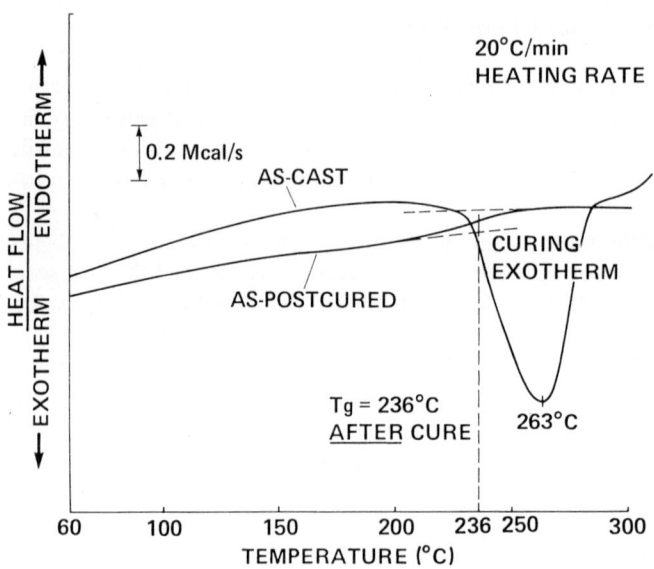

**Fig. 8.** Differential scanning calorimetry traces of as-cast and as-postcured Fiberite 934 epoxy resins

In view of our discussion on the free volume concept, we expect a quenching from above $T_g$ would freeze in a relatively large amount of free volume in epoxies. The temperature of 260 °C for 20 minutes followed by a quenching to 23 °C, we expected to have trapped in a larger amount of free volume compared to the postcured-then-slowly-cooled material. Since more free volume can be interpreted to mean higher chain mobility and shorter molecular relaxation time, we were surprised to find that *in the quenched system*, the toughness was greatly reduced, i.e., both strain-to-break and ultimate tensile strength decreased further compared to the postcured material.

We propose to rationalize the observation by a phenomenon known as residual thermal stresses. Residual thermal stresses arise from the fact that carbon-fiber and epoxy have different thermal expansion coefficients and a quenching of the composite would conceivably produce residual stresses. Apparently, the quenching process may produce enough residual stresses to lower the toughness of the composite. In the absence of such residual stresses the free volume concept alone would predict a quenched glass to have larger amount of free volume and hence constitute a less brittle substance.

In order to study the effect of physical aging on the carbon-fiber reinforced epoxy, the freshly quenched materials were then sub-$T_g$ annealed at 140 °C. After annealing for only 10 minutes at that temperature, the toughness of the composite was restored to a level comparable to that of the postcured material (see Fig. 7). It is likely that residual thermal stresses resulted from the quenching were annealed away during this 10 minutes thermal aging at 140 °C.

Physical aging in the epoxy matrix definitely affects the long-term mechanical properties of the composite. As the material was further sub-$T_g$ annealed at 140 °C for $10^2$, $10^3$, $10^4$, and up to $10^5$ minutes, the toughness and other ultimate properties

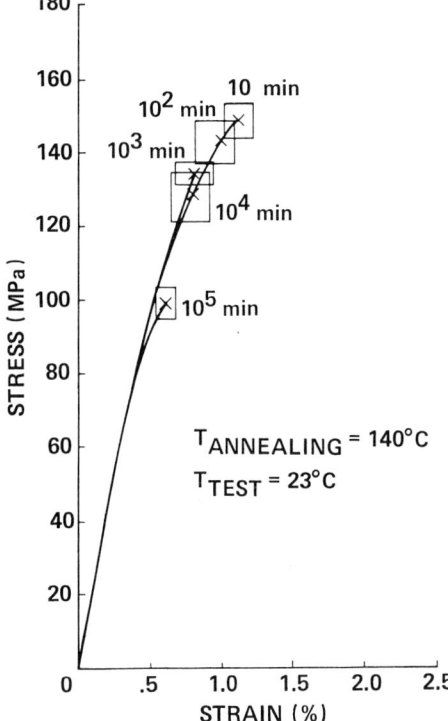

Fig. 9. Effect of sub-$T_g$ annealing on the ultimate mechanical properties of carbon-fiber-reinforced epoxies

continued to decrease. Figures 9, 10, and 11 show convincingly that the decrease in ultimate properties is proportional to the aging time. By $10^5$ minutes at 140 °C, ultimate tensile strength was reduced to 100 ± 3.30 MPa, a net decrease of 38.5% from the original strength of the as-received material. Strain-to-break decreased ca. to 73.6% when one compares the as-received materials to the ones aged for 5

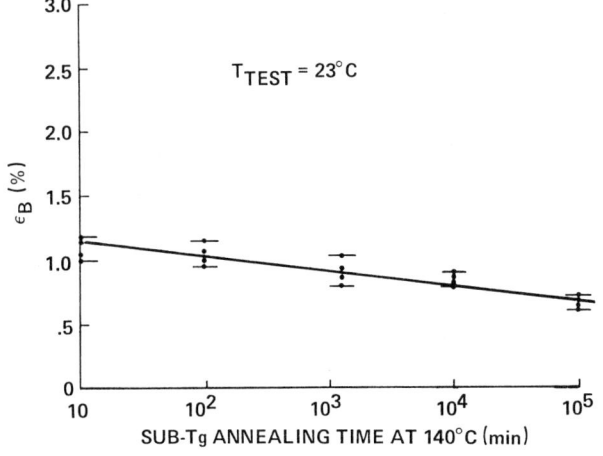

Fig. 10. Strain-to-break measured by axial extensometer as a function of sub-$T_g$ annealing time

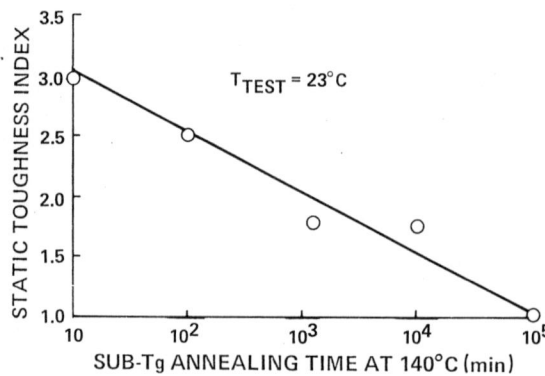

Fig. 11. Normalized toughness index as a function of sub-$T_g$ annealing time

decades ar 140 °C. Toughness, more drastically, decreased to 87.8 % as the composite was allowed to go through the thermal treatment described above and aged up to $10^5$ minutes. For specimens that were aged for 5 decades, we estimate the toughness to have a value of 0.36 Joule cm$^{-3}$.

No weight loss or gain was observed during the sub-$T_g$ annealing experiment. Young's modulus estimation indicated insignificant changes during this series of sub-$T_g$ annealing experiments. Yield stress analysis did not show any change in properties.

The inplane shear stress-strain tests reported here have been well demonstrated to be a reliable test for matrix-dominated properties in composites [141]. For the selected mechanical properties that were monitored, their sensitivity to the thermal history was well demonstrated. In particular, the embrittlement process during the sub-$T_g$ annealing or physical aging has been clearly observed. This decrease in molecular mobility, which gives rise to an increase in relaxation time and hence a decrease in toughness, can be rationalized as a decrease in free volume in an approach towards the equilibrium glassy state.

### 3.3 Stress Relaxation Analysis on Neat Resins

Figure 12 shows the stress relaxation curves of the fully-crosslinked epoxies with 10, $10^2$, $10^3$, $10^4$, and $10^5$ minutes of sub-$T_g$ annealing at 140 °C. As thermal aging progresses, the initial stress level increases and the relaxation rate decreases. The higher initial stress is caused primarily by the higher modulus (as also confirmed by dynamic mechanical data to be discussed in the next section) but is also partially due to less relaxation during the stretching period for aged specimens. The relaxation rate is measured as a percent of stress relaxation during the first 10 minutes of the experiment. Figure 13 shows this percent of stress relaxation as a function of logarithmic aging time for the epoxy resin. The data clearly demonstrate that the rate of stress relaxation decreases with aging time. The thermoreversible nature of physical aging is demonstrated by the generation of similar data obtained for re-aged specimens (Fig. 14). Samples were aged at 140 °C for $10^4$ minutes, annealed above $T_g$ (a "memory-wiping-out" process in which the aging thermal history were erased),

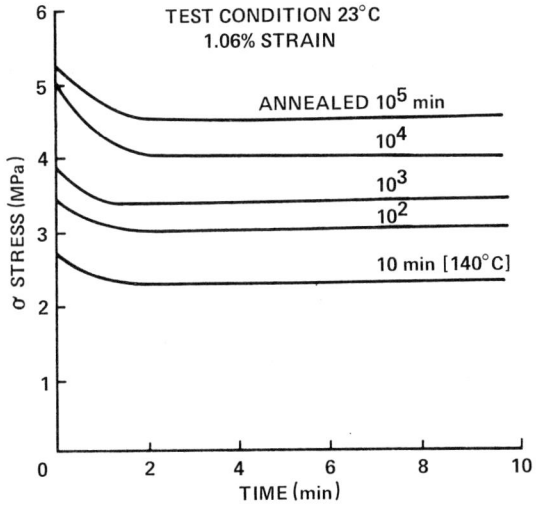

**Fig. 12.** Stress relaxation curves of Fiberite 934 epoxies of different aging histories

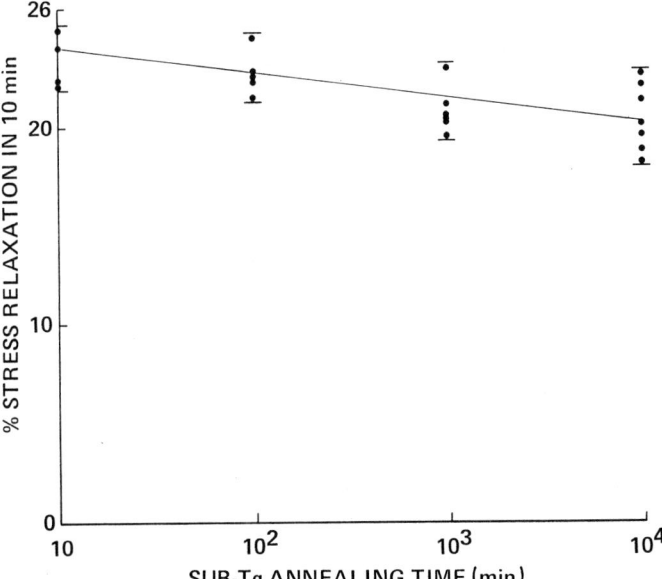

**Fig. 13.** Percent stress relaxation of quenched Fiberite 934 epoxy resins as a function of sub-$T_g$ annealing time

quenched, and then subjected to aging for the second time at 140 °C. Both rounds of aging were conducted in nitrogen and in darkness. The percent of stress relaxation as a function of logarithmic re-aging time for the epoxy resin is shown in Fig. 14.

This time-dependent behavior would be expected based on the model that as network chains lose mobility during the aging process (due to a decrease in free

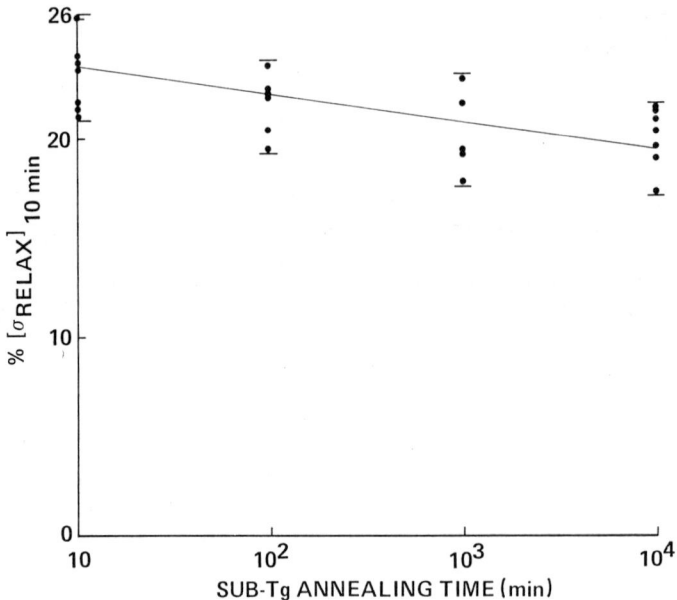

**Fig. 14.** Percent stress relaxation of requenched Fiberite 934 epoxy resins as a function of sub-$T_g$ annealing time

volume), the ability to dissipate stress is reduced. The decrease in stress relaxation rate can therefore be explained on the basis of free volume collapse during sub-$T_g$ annealing.

## 3.4 Dynamic Mechanical Analysis on Composites

Dynamic mechanical analysis of polymeric materials, including epoxies [142-160], is an established tool in measuring the mechanical dispersion peaks as well as other parameters such as the dynamic storage modulus of macromolecules. Wetton has reported some effects by sub-$T_g$ annealing on the modulus and damping peaks of a low-$T_g$ epoxy [161]. In this investigation of high-$T_g$ epoxy-matrix composites, we have observed a decrease in damping storage modulus of the composite as a function of physical aging time. Figure 15 shows a specific example of a 140 °C/$10^2$-minutes-aged carbon/epoxy composite having a glass transition maximum peak temperature near 242 °C. The onset of the $T_g$ is near 175 °C. This composite, which is ±45° carbon-fiber-reinforced, shows a dynamic storage modulus of the epoxy matrix in the glassy-state of ca. 15 GPa. At the onset of the glass-to-rubber transition (see Fig. 15), the modulus drops gradually from 15 GPa (175 °C) to about 3 GPa (300 °C) as the rubbery plateau is reached.

With physical aging at 140 °C in nitrogen/dark atmosphere, the dynamic storage modulus is very sensitive to aging time. The modulus increased from 13 GPa (10 minutes-aged) to 18 GPa for samples aged up to $10^5$ minutes at 140 °C (see Fig. 16). These results agree with observations made in the stress relaxation experi-

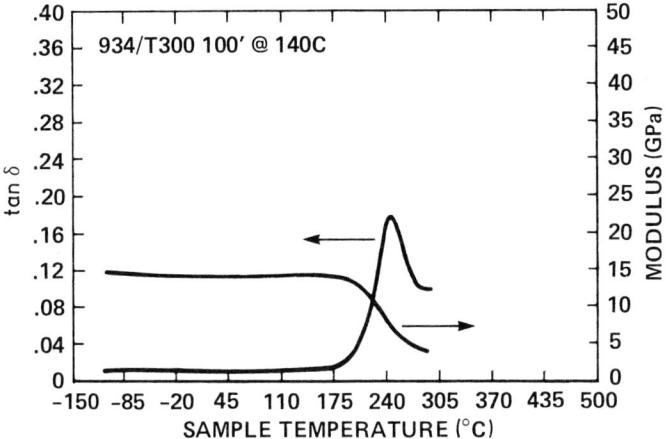

**Fig. 15.** Dynamic mechanical analysis of 100 minutes-aged Thornel 300/Fiberite 934 composite showing the loss tangent and the dynamic storage modulus

ments discussed earlier in this review in which the epoxy tensile modulus increased with sub-$T_g$ annealing.

The mechanical dispersion peaks in low-$T_g$ epoxies such as Bisphenol-A based resin (Epon 828, products from Shell Development Company) have been the subject of numerous studies [143, 145–148, 152–155, 159]. The *alpha*-dispersion peak related to the glass transition can undoubtly be attributed to the large-scale cooperative segmental motion of the macromolecules. The *beta*-relaxation near —55 °C, however, has been the subject of much controversy [146, 153]. One postulated origin of the dispersion peak is the "crankshaft mechanism" at the junction point of the network epoxies (Fig. 17). The "crankshaft motion" for linear macromolecules was first proposed [163–166] as the molecular origin for secondary relaxations which involved restricted motion of the main chain requiring at least 5 and as many as 7 bonds [167]. This kind of

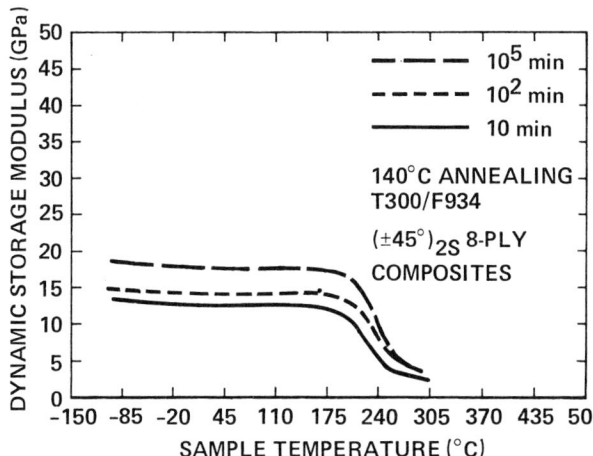

**Fig. 16.** Influence of physical aging time on the dynamic storage modulus of Thornel 300/Fiberite 934 composites

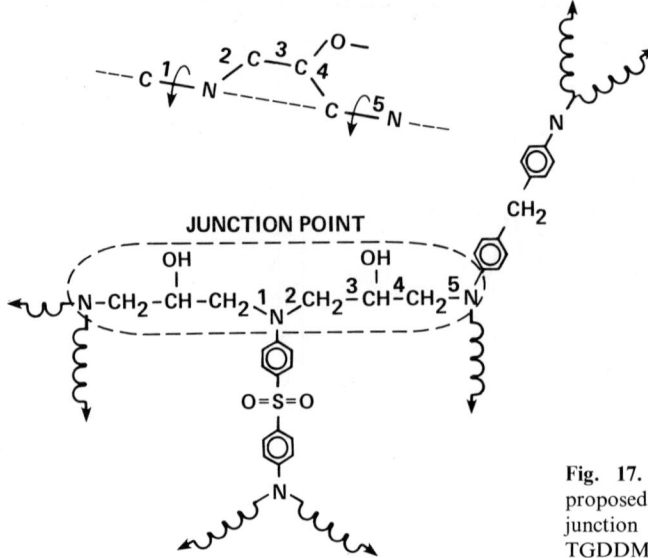

Fig. 17. "Crankshaft motion" as proposed in the literature for the junction point of crosslinked TGDDM-DDS epoxy

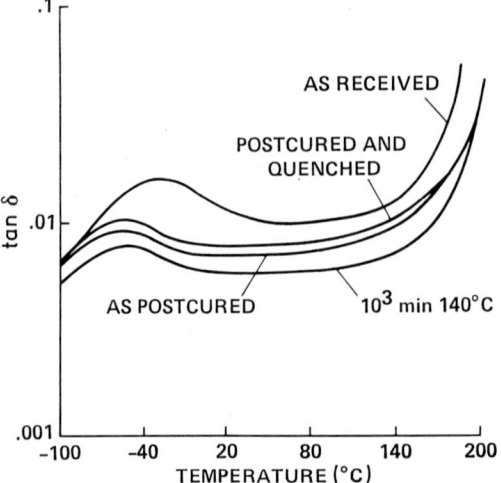

Fig. 18. Secondary mechanical dispersion peaks of Thornel 300/Fiberite 934 composites as influenced by thermal history

crankshaft rotation needs an energy of activation of the order of 11 to 15 kcal/mol and most likely requires creation of free volume in order that the "crankshaft" may rotate [167].

Figure 18 shows the effects of thermal history on the mechanical dispersion peaks in Fiberite 934 epoxy composites. The as-fabricated materials show by far the largest damping which spans from $-100\,°C$ to $20\,°C$. Postcuring completes the crosslinking and results in significantly lower-damping material.

Physical aging affects significantly the mechanical damping in both the *alpha* and *beta* relaxation peaks. Figure 19 shows the decrease in the *beta* loss peak as a func-

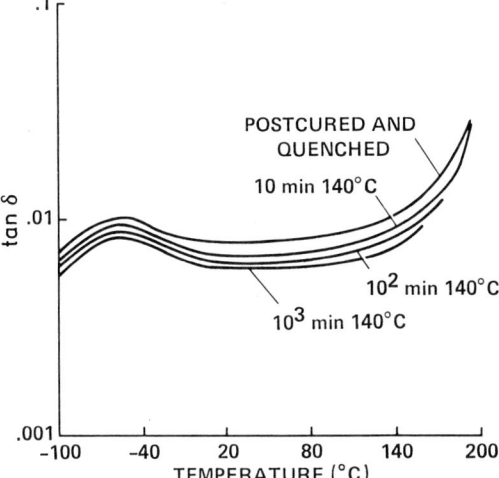

**Fig. 19.** The influence of physical aging time on the secondary loss peak of Thornel 300/Fiberite 934 composites

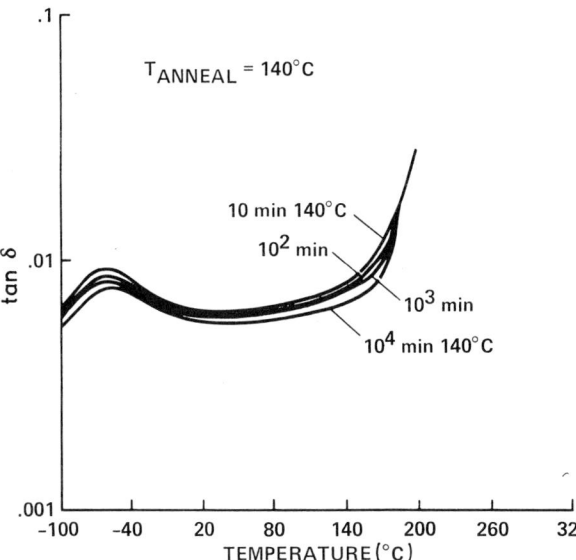

**Fig. 20.** Influence of requenching followed by reaging on the secondary loss peak of Thornel 300/Fiberite 934 composites

tion of aging time at 140 °C in nitrogen. This gradual decrease in damping can be explained by a relaxation in which the epoxy network loses mobility and free volume during its asymptotic approach towards the equilibrium glassy state; as a result, the ability to dissipate energy is reduced. This is a significant observation in view of the fact that the area under the secondary mechanical dispersion peak is often correlated with the impact resistance of the polymer [168]. Upon requenching from above $T_g$ and re-aging such material, the thermoreversible nature of physical aging can be demonstratively shown. The effect of 140 °C aging on the *beta*-transition in the epoxy matrix of requenched specimens is shown in Fig. 20.

With sub-$T_g$ annealing at 140 °C the maximum peak temperature of $T_g$ tended to shift to higher temperatures. For example, the value was 242.0 °C for 10 minutes-aged samples. In the two aging/reaging experiments, $T_g$ shifted to 253.0 °C for both $10^5$ minutes-aged and reaged samples. $T_{beta}$, however, appeared to be less sensitive to physical aging time and the *beta* maximum peak temperature stayed at about −55 °C.

## 3.5 Differential Scanning Calorimetry on Neat Resins

DSC was utilized to study both the state of the cure (extent of crosslinking) and the kinetics of enthalpy relaxation in crosslinked epoxies [18,21,56,71,72]. DSC results confirmed a fully crosslinked epoxy network having no exotherm at temperatures up to 300 °C. As shown in Fig. 8, an exotherm with a maximum curing peak temperature at 263 °C was detected during the scan from about 60 to 300 °C. Because continued crosslinking can change the epoxy properties, it was important to this study that all chemical aging be eliminated to isolate the physical aging effects. Therefore, the epoxies were postcured at 250 °C for 16 hours in nitrogen. After the postcuring, DSC of the epoxy network showed a step-function increase in heat capacity during the $T_g$ transition from 180 to 270 °C (Fig. 8), which indicated a fully cured state for the thermoset. In addition, FTIR studies, monitoring the 910 cm$^{-1}$ epoxy three-membered ring absorption band, indicated that, with our postcuring schedule, all epoxide rings were opened during the polymerization. The disappearance of the 910 cm$^{-1}$ band as a monitor of curing is a standard practice in assuring complete cross-linking [169].

DSC has been amply demonstrated in the pat to be a useful instrument to follow the kinetics of enthalpy relaxation in polymers [41,42,74]. Figure 21 shows the DSC

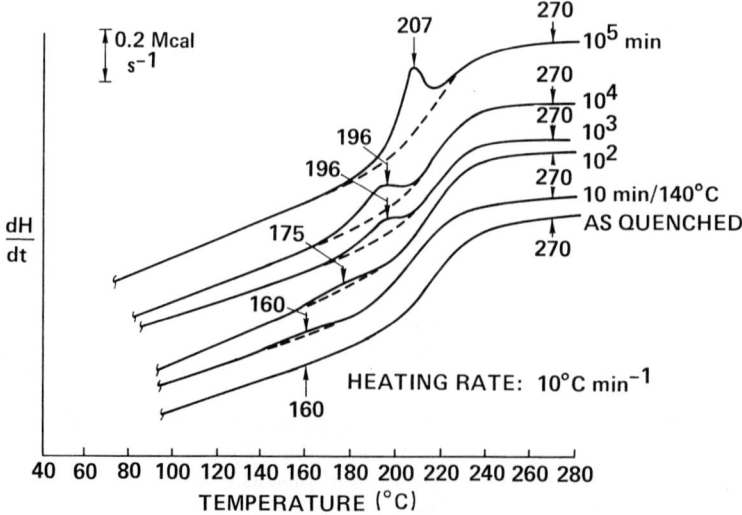

**Fig. 21.** Influence of physical aging on the endothermic enthalpy relaxation of Fiberite 934 resin

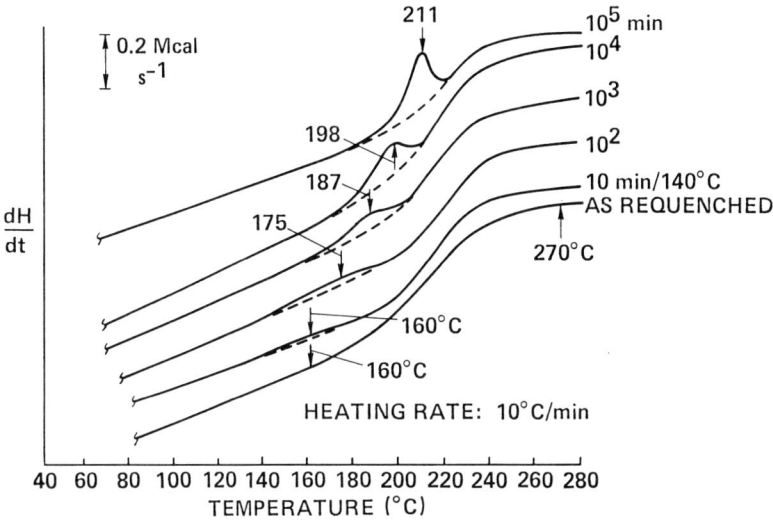

**Fig. 22.** Effect of reaging on the enthalpy relaxation of requenched Fiberite 934 resin

scans of the fully-crosslinked epoxy specimens which were quenched from above $T_g$ and then subjected to aging at 140 °C. The full line is the first scan while the dotted line represents the second scan taken right after rapid cooling from the initial one. The following observations were made:
1. The enthalpy relaxation peak appears near the onset of the transition from the glassy state to the rubbery state. This peak appears after only 10 minutes of aging.
2. During sub-$T_g$ annealing, the relaxation peak shifts to higher temperature and grows in magnitude.
3. This recovery phenomenon is thermoreversible. Upon re-aging the material which was cooled from above $T_g$, the relaxation peak will reappear and grow with time. (See Fig. 22.)

Enthalpy relaxation studies were carried out at 3 different temperatures: 80, 110, and 140 °C. Compared to the 140 °C enthalpy relaxation data, the 110 °C aging

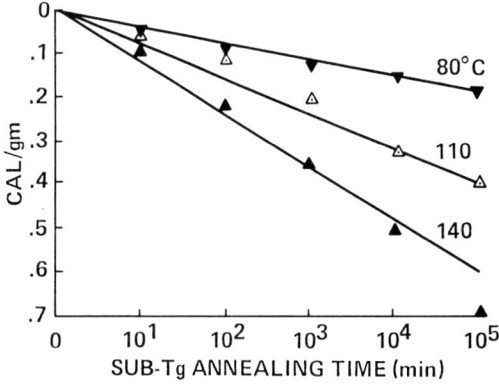

**Fig. 23.** Enthalpy loss at 3 different temperatures of aging as a function of log sub-$T_g$ annealing time

kinetics are definitely slower. Even though "aging peaks" were again observed for 80 °C annealing. This time, the relaxation peak was much smaller compared to the 110 °C aging data. During 80 °C aging, the peak temperature shifted from 100 °C (10 minutes) to 125 °C ($10^5$ minutes). In the case of 110 °C aging, the peak temperature shifted from 130 °C to 180 °C (10 minutes to $10^5$ minutes). In the case of 140 °C aging, we noticed a shift of the relaxation peak from 160 °C (10 minutes) to 210 °C ($10^5$ minutes).

As mentioned earlier, the relaxation enthalpy was measured by superimposing the first and the second DSC scans for each specimen. Figure 23 shows the relaxation-enthalpy loss versus logarithmic sub-$T_g$ annealing time at 80°, 110°, and 140 °C. There is clearly a linear relationship between the enthalpy relaxation process and the logarithmic aging time, $T_a$.

Figure 23 demonstrates that aging kinetics slow down as the temperature increment ($T_g - T_a$) increases, i.e., the recovery process is a thermally stimulated phenomenon which requires segmental mobility of the polymer in its glassy state.

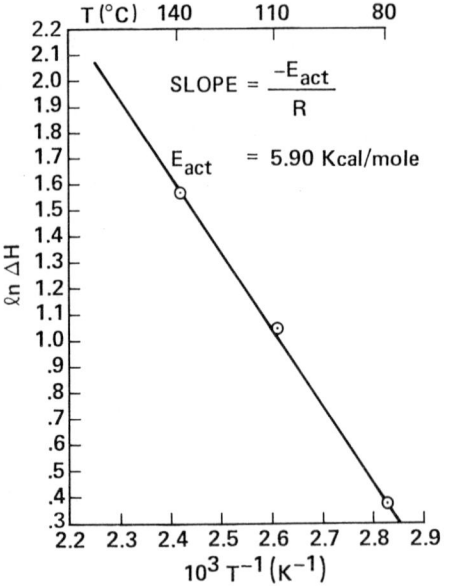

Fig. 24. Arrhenius analysis of the enthalpy loss data of Fig. 23

The lower the sub-$T_g$ annealing temperature, $T_a$, the slower is the aging kinetics. Since there exists a linear relationship between enthalpy loss and aging time, we can determine the activation energy of the enthalpy relaxation process assuming the Arrhenius equation holds. Figure 24 shows the Arrhenius plot. From the slope, we can estimate the activation energy to be 5.9 kcal/mol. This activation energy is very close to the typical hydrogen bond dissociation energy for a majority of hydrogen-bonded systems [170,171]. It is suggested that during the resin contraction or densification process during volume relaxations, hydrogen bonds may be broken and re-formed.

The activation energy for epoxy polymer relaxation of 5.9 kcal/mol estimated from the Arrhenius analysis appears to be low value compared to enthalpy relaxation in inorganic glasses such as the $B_2O_3$ system reported by Moynihan et al.[172]. Other researchers, however, have also noted an activation energy for a low-$T_g$ epoxy to be ca. 6 kcal/mol[173]. This apparent discrepancy between macromolecular and inorganic systems may stem from a difference in relaxation mechanisms in those two classes of glassy materials.

## 3.6 Thermal Mechanical Analysis on Neat Resins

Thermal mechanical analysis was utilized by Ophir[174] to study the densification of Bisphenol-A-based epoxies. The glass transition temperature can easily be characterized by a slope change as the resin transits from the glassy state to the rubbery state (see Fig. 25). Hence, in glassy material, it is typically represented by two thermal expansivity parameters, one below $T_g$ (glassy thermal expansivity) and one above $T_g$ (rubbery thermal expansivity).

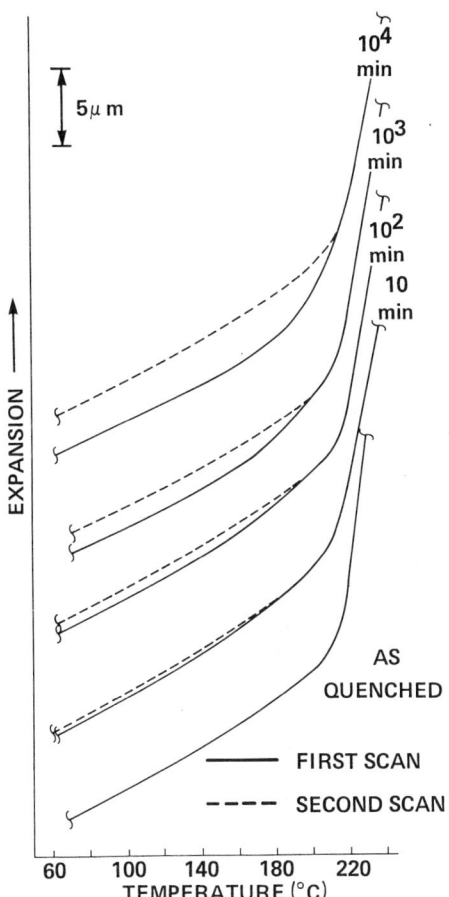

Fig. 25. Thermal expansion behavior of Fiberite 934 epoxies as influenced by aging history

Figure 25 shows the thermal expansion behavior of a fully-crosslinked epoxy as a function of aging time at 140 °C sub-$T_g$ annealing. By superimposing the first scan (for aged material) and the second scan (for as-quenched material) at the high-temperature rubbery region, it is possible to monitor the development of aging in the resin, i.e., the progress of the densification process.

As shown in Fig. 25, the aged glass typically has a lesser volume in the glassy state compared to the as-quenched state. It is obvious from the data that the longer the aging time, the larger is the amount of volume lost due to sub-$T_g$ annealing. This observation also fits well into the "free-volume collapse" model discussed earlier.

Figure 26 shows the thermal expansion behavior of requenched epoxies. Upon reaging, the densification process was again measurable. Data shown in Figure 26, therefore, supports the notion that physical aging processes are thermoreversible.

By analyzing the linear portion of the thermal expansion curves below and above $T_g$, it is possible to calculate the expansivity of each specimen taking into account its individual thickness. Through such analysis, significant variations were observed in the thermal expansivity of the cured epoxy both below and above its $T_g$.

Figure 27 shows the expansivity variations as a function of thermal history. As-cast epoxy has a value of $5.43 \times 10^{-5}$ K$^{-1}$, (below $T_g$). Expansivity below $T_g$ decreased

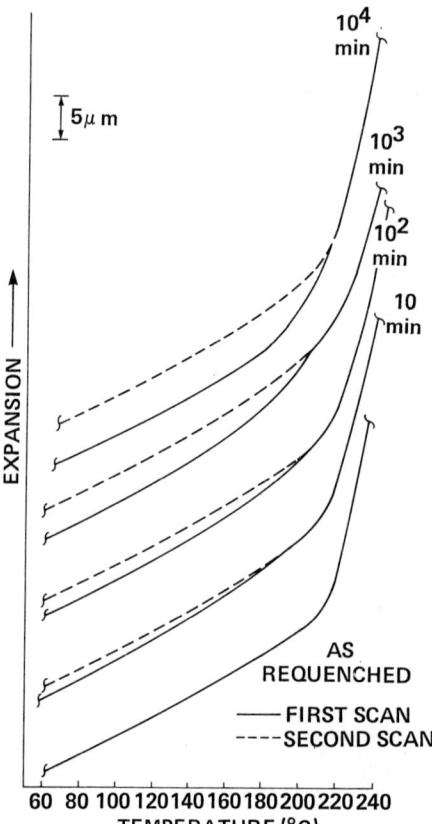

Fig. 26. Thermal expansion behavior of Fiberite 934 epoxies as influenced by requenching (erasure of thermal history) and subsequent reaging

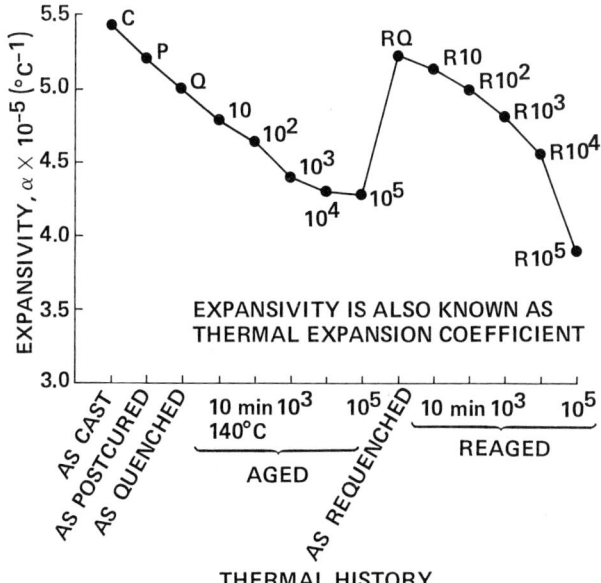

Fig. 27. Glassy state thermal expansivity of Fiberite 934 epoxies as a function of thermal history

with postcuring to $5.20 \times 10^{-5}$ K$^{-1}$, which is reasonable because postcuring resulted in a network which has a higher crosslink density and hence, lesser mobility. Quenching, which introduced a thermal shock and also residual thermal stresses, caused the epoxy to be less expansible below $T_g$ ($4.98 \times 10^{-5}$ K$^{-1}$) in spite of the increased free volume through quenching. In this experiment, similar to the results suggested by the tensile stress-strain analysis, residual thermal stresses seem to override the importance of free volume considerations in affecting the glassy expansivity of as-quenched resin.

With 10 minutes of sub-$T_g$ annealing at 140 °C, thermal expansivity below $T_g$ decreased to $4.78 \times 10^{-5}$ K$^{-1}$. This parameter decreased throughout the 140 °C aging experiment. After $10^5$ minutes of aging, the value decreased to $4.30 \times 10^{-5}$ K$^{-1}$. The free volume decrease evidently dictates the thermal expansivity in the glassy state during sub-$T_g$ annealing.

To erase resin history, aged samples were quenched from above $T_g$. With the requenching, glassy state expansivity was restored to a high value of $5.22 \times 10^{-5}$ K$^{-1}$, comparable to the as-postcured value. Requenching obviously has introduced a significantly larger amount of free volume in the resin and thus made the resin more expansible (see Fig. 27).

Reaging the resin at 140 °C again caused a decrease in glassy-state expansivity buth the decrease was larger than the first round of aging. Evaluating the decrease for $10^4$ minutes data in the two series of aging indicated that the aging kinetics was similar.

Figure 28 shows the thermal expansivity of epoxy above its $T_g$ as a function of thermal history. Rubbery-state expansivity is generally an order of magnitude larger compared to the glassy-state expansivity (see Table 2). As-cast epoxy has an expansivity above $T_g$ of $3.22 \times 10^{-4}$ K$^{-1}$. With postcuring and quenching, this parameter tends

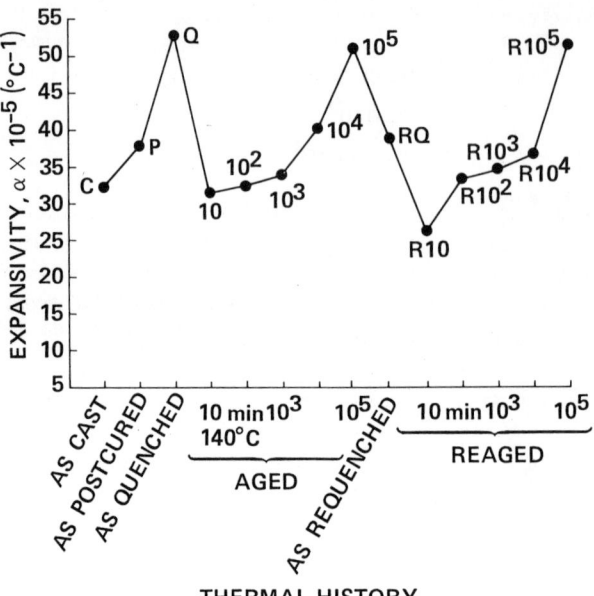

**Fig. 28.** Rubbery state thermal expansivity of Fiberite 934 as a function of thermal history

| THERMAL HISTORY | | $\alpha$(BELOW $T_g$) × $10^5$ °C$^{-1}$ (ca. 60 – 160°C) | | $\alpha$(ABOVE $T_g$) × $10^5$ °C$^{-1}$ (ca. 200 – 240°C) | |
|---|---|---|---|---|---|
| AS-CAST | | 5.43 | | 32.2 | |
| AS-POSTCURED | | 5.20 | | 37.7 | |
| AS-QUENCHED | | 4.98 | | 52.9 | |
| 140°C AGED | 10 | 4.78 | DE-CREASE WITH AGING | 31.4 | IN-CREASE WITH AGING |
|  | $10^2$ | 4.63 | | 32.4 | |
|  | $10^3$ | 4.40 | | 33.9 | |
|  | $10^4$ | 4.32 | | 40.1 | |
|  | $10^5$ min | 4.30 | | 51.1 | |
| AS-REQUENCHED | | 5.22 | | 38.8 | |
| 140°C REAGED | 10 | 5.13 | DE-CREASE WITH AGING | 26.2 | IN-CREASE WITH AGING |
|  | $10^2$ | 5.00 | | 33.3 | |
|  | $10^3$ | 4.82 | | 34.5 | |
|  | $10^4$ | 4.56 | | 36.6 | |
|  | $10^5$ min | 3.88 | | 51.5 | |

to increase due to the interplay of free volume variations and factors involving residual thermal stresses.

As is clearly indicated by the data in Figure 28, expansivity above $T_g$ for Fiberite 934 epoxy tend to increase with aging at 140 °C. This increase in expansivity in the rubbery state can be traced to a "catching-up-process" for volume lost during physical aging. At temperatures above $T_g$, there is enough thermal energy for the system to reach equilibrium. Since volume is lost during physical aging, the high temperatures provide the thermal energy to recover for the "lost volume". The longer the glass is subjected to aging, the higher would be its tendency to recover the lost volume, hence manifested in a larger value of thermal expansivity above $T_g$. An increase from $3.14 \times 10^{-4}$ K$^{-1}$ to $5.11 \times 10^{-4}$ K$^{-1}$ (62.7% increase) was observed.

With requenching and reaging, expansivity above $T_g$ again increases with reaging time (96.6% increase), demonstrating once again the thermoreversibility of physical aging. Table 2 summarizes the results of the thermal mechanical analysis.

## 3.7 Density Measurements on Neat Resins

The density of the cured epoxy was followed as a function of its thermal history. Postcuring caused the largest decrease in density, from 1.290 g cm$^{-3}$ for as-cast epoxy to 1.230 g cm$^{-3}$ for as-postcured/slowly cooled epoxy. The decrease in room-temperature density can be explained partially by escape of unreacted DDS crosslinker (density of DDS = 1.380 g cm$^{-3}$) and/or other low molecular weight species during postcuring. Aherne et al. [175] also observed a decrease in density with Epon 828 epoxy postcuring, but argued from the point of view of free volume to explain the observations.

A priori, one would assume on the basis of crosslink density that a postcured system (presumably with a higher crosslink density) would have a higher density. This may actually be the case at the postcure temperature. At room temperature, however, Gillham et al. [176] have indicated that the higher-crosslinked system may be quenched further from the hypothetical equilibrium glassy state, resulting in a higher value of free volume or lower density.

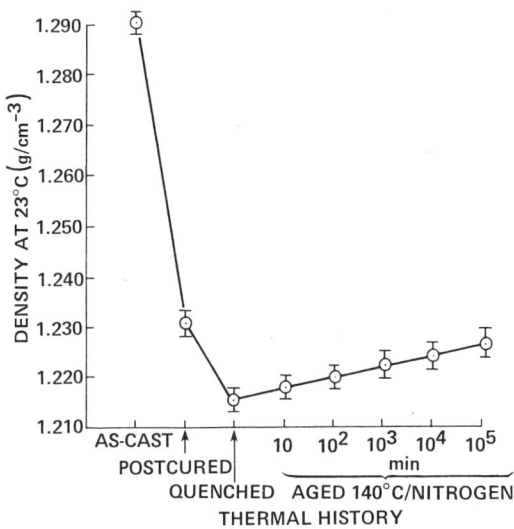

Fig. 29. Density of Fiberite 934 epoxy as a function of thermal history

With an air-quench, the density of the fully-crosslinked epoxy drops from 1.230 to 1.215 g cm$^{-3}$. With sub-$T_g$ annealing, an increase of 0.82% in the resin was observed during the 140 °C aging. This direct piece of evidence agrees perfectly with the "free volume collapse" model in which the resin densifies. Figure 29 summarizes these observations.

## 3.8 Hardness Measurements on the Neat Resins

The hardness of a material is related to its resistance to scratching or denting. The hardness value for the cured epoxy is very dependent on its thermal history and is also a time-dependent parameter during physical aging. M-scale of the Rockwell Hardness Index is reported. As cast epoxy has a low hardness value of M70 ± 3. With postcuring, the value increased to M97 ± 3. This represents a 39% increase of hardness with postcuring. With an air-quench, the epoxy hardness dropped slightly for about 4% to M93 ± 3. This is understandable since the excess trapped free volume in the as-quenched epoxy may well soften the system. With aging at 140 °C, the epoxy hardened from M93 ± 3 to M110 ± 3 for $10^4$ minutes-aged material (an 18% increase). With a requenching of 10 minutes-aged epoxy, a drop of 7% in the hardness is observed, likely due to a restoration of free volume in the resin, as manifested in a softened hardness index of M92.5 ± 2.5. Hence, once again, the thermoreversibility of the physical aging process is demonstrated. Figure 30 summarizes the observations.

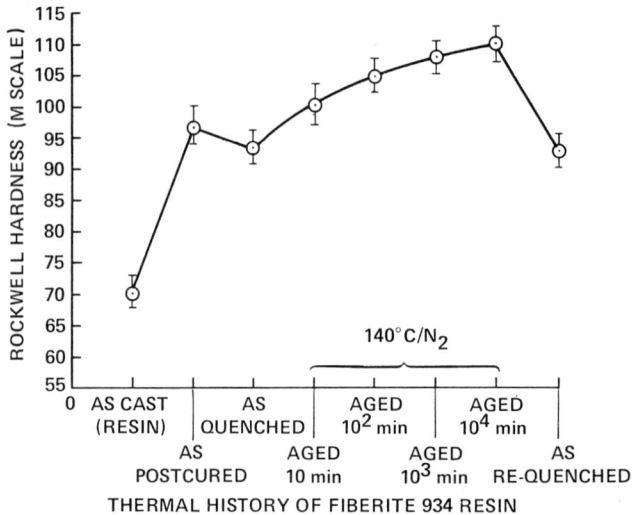

**Fig. 30.** Hardness of neat Fiberite 934 epoxy as a function of thermal history

## 3.9 Moisture Sorption Kinetics on Neat Resins

Numerous studies have been reported on the effect of diffusion of water into epoxy resins [177,178]. It is generally agreed that moisture acts as a plasticizer which lowers the $T_g$ of the resin [130–134]. Very little work, however, has been reported on the effect of physical aging on the diffusion behavior of water into network epoxies. This Section of the review summarizes the first attempt to study such an effect on TGDDM-DDS epoxy/water interactions.

Epoxy glasses aged at 140 °C were subjected to 40 °C/98% relative humidity moisture penetration. Figure 31 shows the results of this transport experiment. We observed both a decrease of initial sorption kinetics as well as a decrease of equilibrium sorption level as a function of aging time. This supports the idea that during sub-$T_g$ annealing, the resin contrasts and densifies, resulting in decreased free volume.

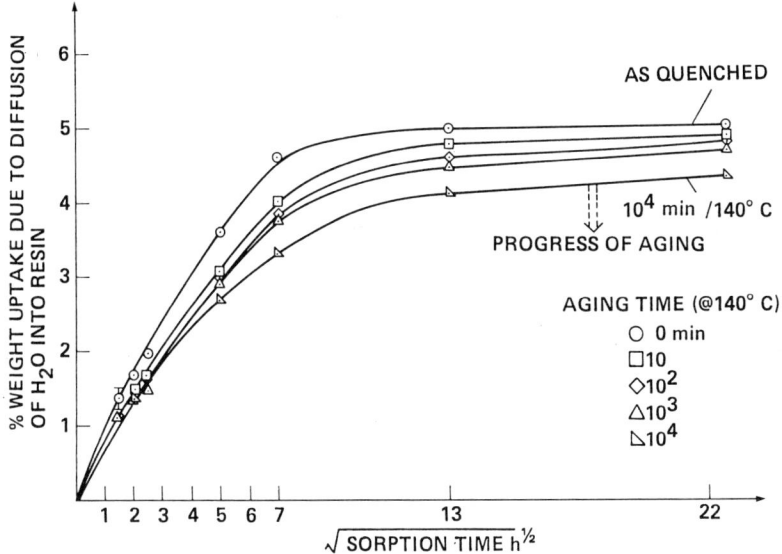

**Fig. 31.** Moisture absorption behavior of fully crosslinked Fiberite 934 resins as influenced by physical aging

In another series of diffusion experiments, as-postcured epoxies were first immersed in 23 °C heavy water for 2 months. Then the temperature of the epoxy/heavy water interacting system was increased to 40 °C. The continuous influx of heavy water into the epoxy water can be easily monitored by deuterium NMR spectroscopy. The free-induction-decay signal as normalized by the specimen weight showed an increase as a function of sorption time (see Fig. 32). For example, a two-months room-temperature sorption resulted in an epoxy with 2.10% moisture as determined by gravimetry. With 667 hours of additional sorption at 40 °C, 3.14% of moisture resided in the epoxy. Correspondingly, the sorption kinetics of NMR free-induction decay signal showed an increase from 275 (arbitrary units) for 2-months/23 °C diffusion to 990 units with additional 677 h of diffusion at 40 °C (Fig. 32). In general, the epoxy-heavy water NMR data agree well with that from the epoxy/water gravimetry experiment.

In addition to detecting the heavy water content, we also utilized the deuterium NMR technique to study moisture/epoxy interactions. Figure 33 shows the 4.65 ppm nuclear magnetic resonance of isotopically tumbling heavy water molecules. Figure 34 shows the NMR spectrum of deuterium oxide as the moisture analog resides in an

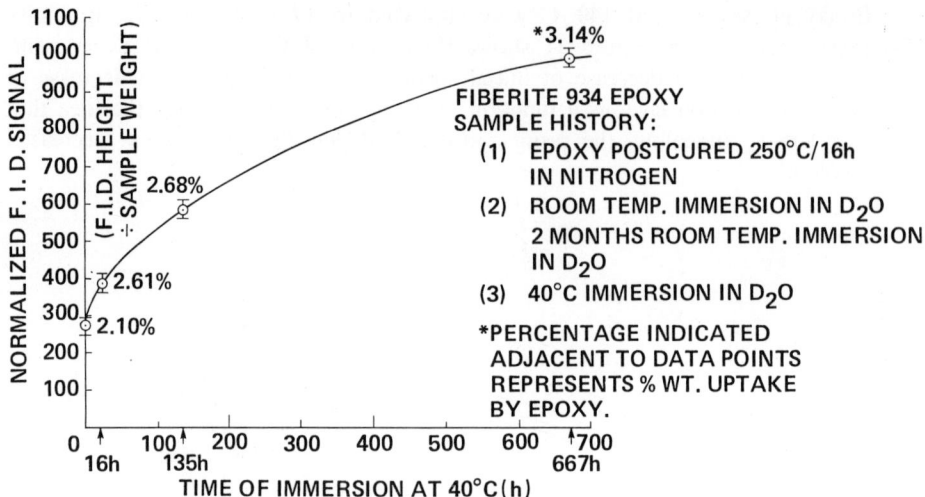

Fig. 32. Heavy water absorption by Fiberite 934 as a function of immersion time at 40 °C as monitored by deuterium NMR spectroscopy

epoxy which had been quenched and stored in heavy water for 2 months at 23 °C and then 1 month at 40 °C.

In Figure 34, we can clearly see the sharp component with little broadening that is due to isotopically tumbling or less-bound heavy water. The wide broadening

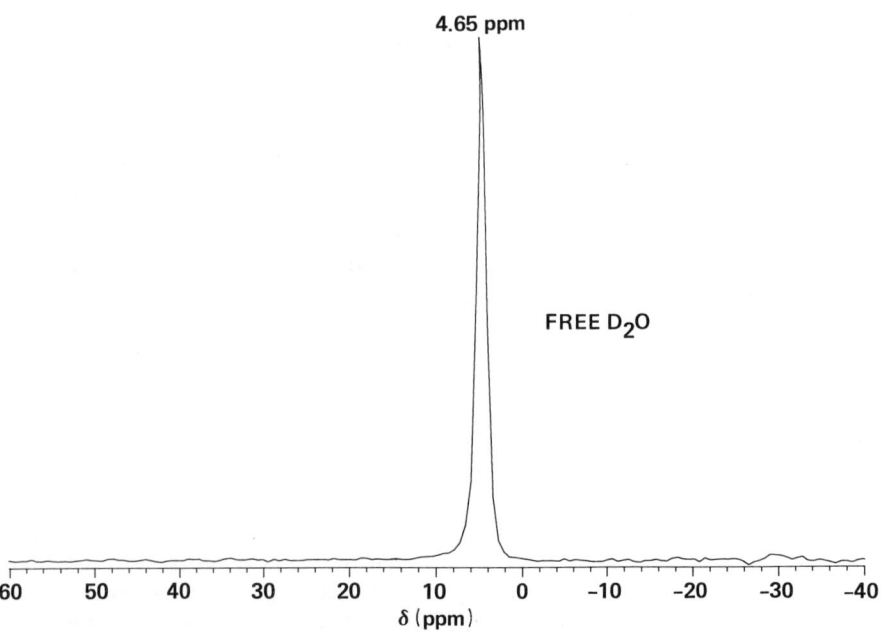

Fig. 33. Deuterium NMR spectrum of isotopically tumbling heavy water molecules

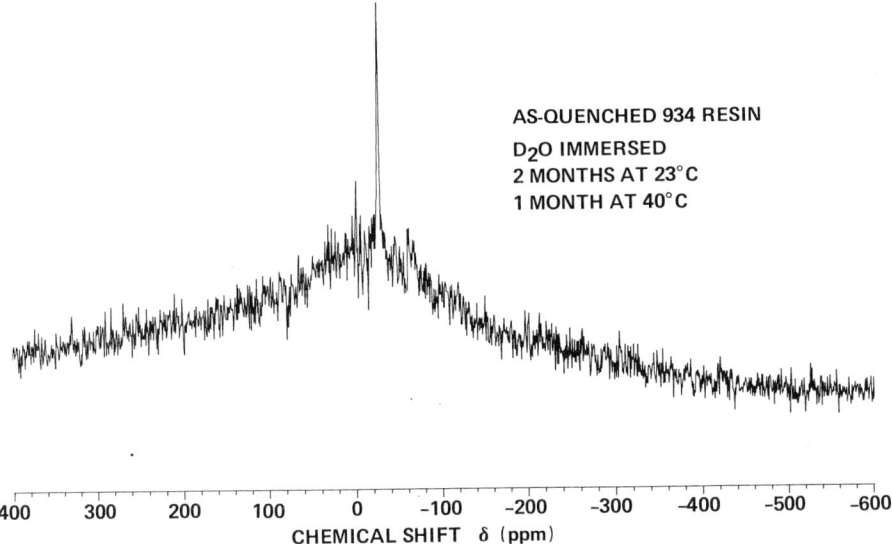

Fig. 34. Deuterium NMR spectrum of heavy water absorbed by a quenched Fiberite 934 epoxy

of the deuterium resonance could be due to deuterium oxide interacting or trapped by the hydrogen bonds of epoxy resin. It is possible that the deuterons of heavy water may exchange with protons in the epoxy (177) so the resonance broadening may only reflect deuterons that have exchanged with protons and now reside as part of the epoxy network. More experiments with deuterated resin can clarify this point.

In the TGDDM-DDS epoxies, hydrogen bonds may be formed among the polar groups. Figure 35 summarizes such hydrogen bonding possibilities. Similar to ordinary water, it is well-known that heavy water is also highly associated due to hydrogen bonding [170, 171]. It is highly likely that the hydrogen-bonded heavy water can disrupt the epoxy hydrogen bonds, thereby causing swelling in the moisture-penetrated resin network. This mode of interactions are shown in Fig. 36. On the other hand, there exist voids (including free volume) in the resin in which the moisture molecules can tumble in a less-bound fashion. We therfore propose a "Dual Mode of Sorption" model [189] for this epoxy/moisture system: some less-bound moisture molecules can tumble in a less-bound fashion. We therefore propose a giving rise to the sharp NMR component, whereas some moisture could be trapped by the epoxy hydrogen bonds resulting in the broadened NMR component. From the broadened resonance linewidth, it is estimated that the correlation time for heavy water among the polymer hydrogen bonds is in the order of $10^{-7}$ s.

Swelling efficiency is a term used to describe the degree of polymer-solvent interactions [137]. In this investigation, the swelling efficiency data are reported in Fig. 37 where the change in volume of this resin as moisture is absorbed versus the volume of water that has diffused into the network is plotted. The data have been normalized as per cent of the dry weight of the polymer. The data clearly indicate that the swelling efficiency increased as the resin was subjected longer duration of aging.

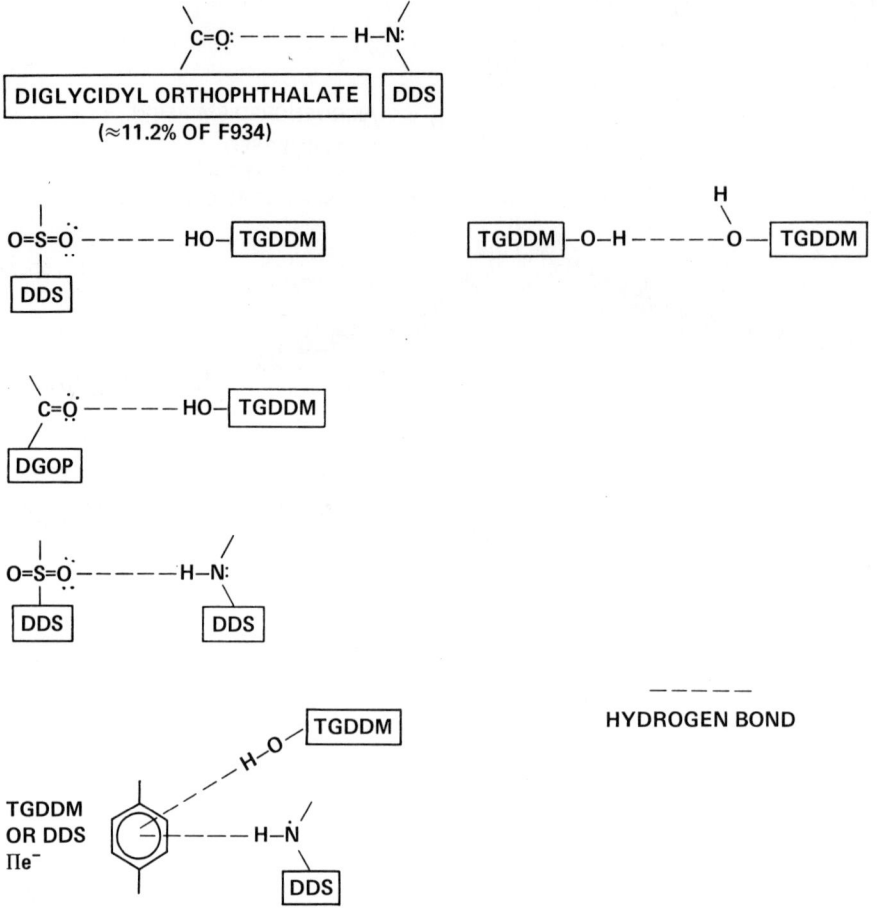

**Fig. 35.** Various hydrogen bonding possibilities by polar groups in Fiberite 934 epoxy. Note that diglycidyl orthophthalate is given the acronym DGOP

Apparently, there is less free volume in the aged polymer network. Any water coming into the aged network would tend to swell the polymer because there are simply less vacant sites in the distribution of free volume. In short, there is more polymer-solvent interactions as water diffuses into an aged epoxy network.

The epoxy-water interactions during the diffusion experiment can be conveniently measured by determining the diffusivity as a function of the aging time. The details of calculating diffusivity in a polymer network have been reported earlier in the literature [179]. In this investigation, diffusivity has been observed to decrease as a function of aging time at 140 °C. Prior to aging, diffusivity was measured to be $5.18 \times 10^{-13}$ m$^2$ s$^{-1}$ at the temperature of 40 °C. With just 10 minutes of sub-$T_g$ annealing at 140 °C, diffusivity decreased to $3.26 \times 10^{-13}$ m$^2$ s$^{-1}$. With further aging, diffusivity changed from $3.26 \times 10^{-13}$ m$^2$ s$^{-1}$ to $2.67 \times 10^{-13}$ m$^2$ s$^{-1}$ after $10^2$ minutes of aging. This parameter decreased further to $2.64 \times 10^{-13}$ m$^2$ s$^{-1}$ after aging $10^3$ minutes and then stayed at that value up to $10^4$ minutes of aging.

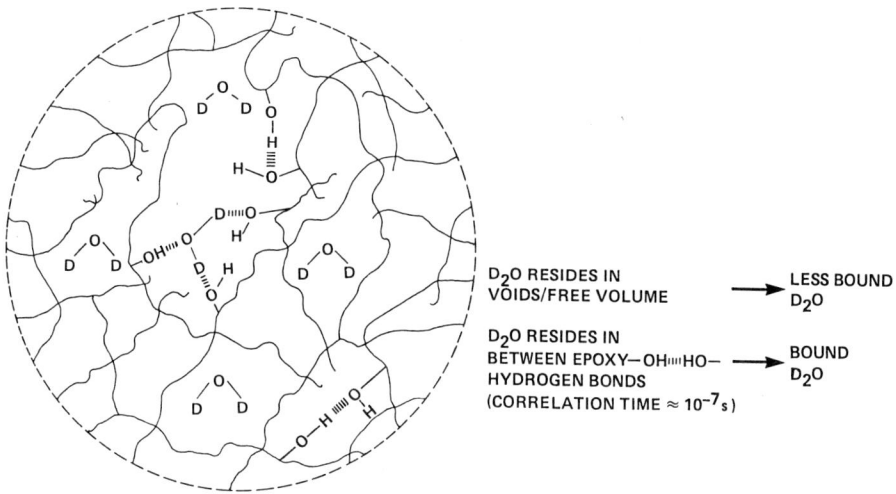

**Fig. 36.** Proposed model for "heavy water/epoxy interactions": heavy water molecules may form aggregates in the free volume of the polymer or disrupt the hydrogen bonds in the resin

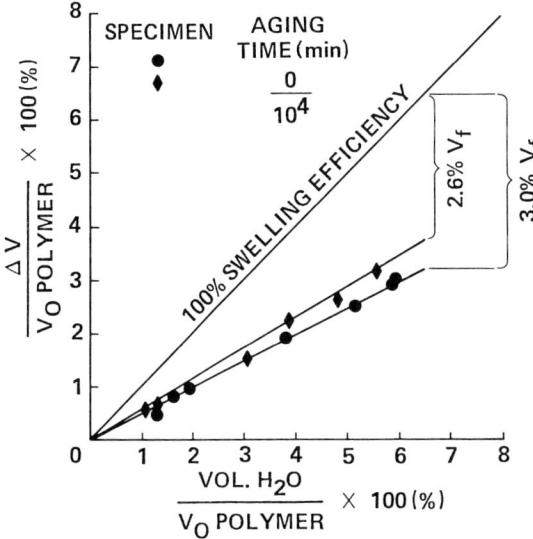

**Fig. 37.** Swelling efficiency and free volume estimation in well-cured Fiberite 934 network epoxies, as affected by sub-$T_g$ annealing. Percent swelling is plotted on the ordinate axis while percent of moisture uptake is plotted on the abscissa

It was possible to estimate the amount of free volume changes on a quantitative basis using the swelling efficiency data as shown in Fig. 37. The free volume in the resin is taken as the difference along the ordinate between the theoretical 100% swelling efficiency line and the experimental swelling efficiency line, after the equilibrium moisture sorption level has been reached. The basis for the quantitative estimation has been reported elsewhere [137]. We observed that free volume in the network resin decreased from ca. 3.0% prior to aging to ca. 2.6% after aging for $10^4$ minutes at 140 °C.

## 3.10 Proton-decoupled CP/MAS NMR Studies on Neat Resins

Proton decoupling, cross polarization, and magic angle spinning techniques were applied for the first time to study the TGDDM-DDS epoxy. Figure 38 shows a carbon-13 NMR spectrum of an as-cast epoxy. In this spectrum, the aromatic carbons (residing in downfield between 100 and 170 ppm) can clearly be resolved from the aliphatic carbons (between 20 and 80 ppm). By integration, the population of aliphatic and aromatic carbons was shown to be roughly the same.

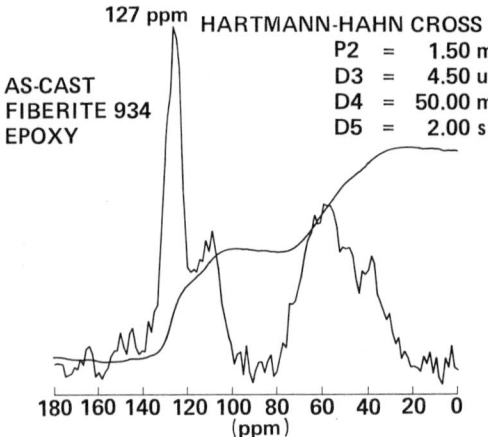

Fig. 38. Proton-decoupled CP/MAS carbon-13 NMR spectrum of an as-cast Fiberite 934 epoxy showing the integration of aromatic and aliphatic carbons

In order to study physical aging at the molecular level, CP/MAS NMR was utilized to study the epoxy densification process. Figure 39 shows the spectral difference between epoxies as cast and after 10 and $10^5$ minutes of aging. Postcuring and aging clearly have resulted in many spectral changes. For example, a resonance peak at ca. 117 ppm for as-cast epoxies disappeared after the completion of chemical crosslinking during the postcure schedule. Of greatest interest, however, was the observation that the sharp and highest aromatic resonance at 127 ppm (in as-cast epoxy) tended to shift downfield with sub-$T_g$ annealing. With aging to $10^5$ minutes, this resonance peak shifted to about 131 ppm. One can interpret this downfield shift as a manifestation of aggregation among aromatic moieties in the resin causing "ring-current effects" [180] during the volume recovery process.

Such aggregation of the phenyl moieties of TGDDM-DDS epoxies during sub-$T_g$ annealing agrees well with the densification model for physical aging. Indeed, it has reported earlier in this review that a 0.82% increase in bulk density was observed during 2 months of 140 °C aging for this epoxy. The spectroscopic evidence presented here shows the elegance of using a molecular probe to substantiate the densification model for physical aging processes.

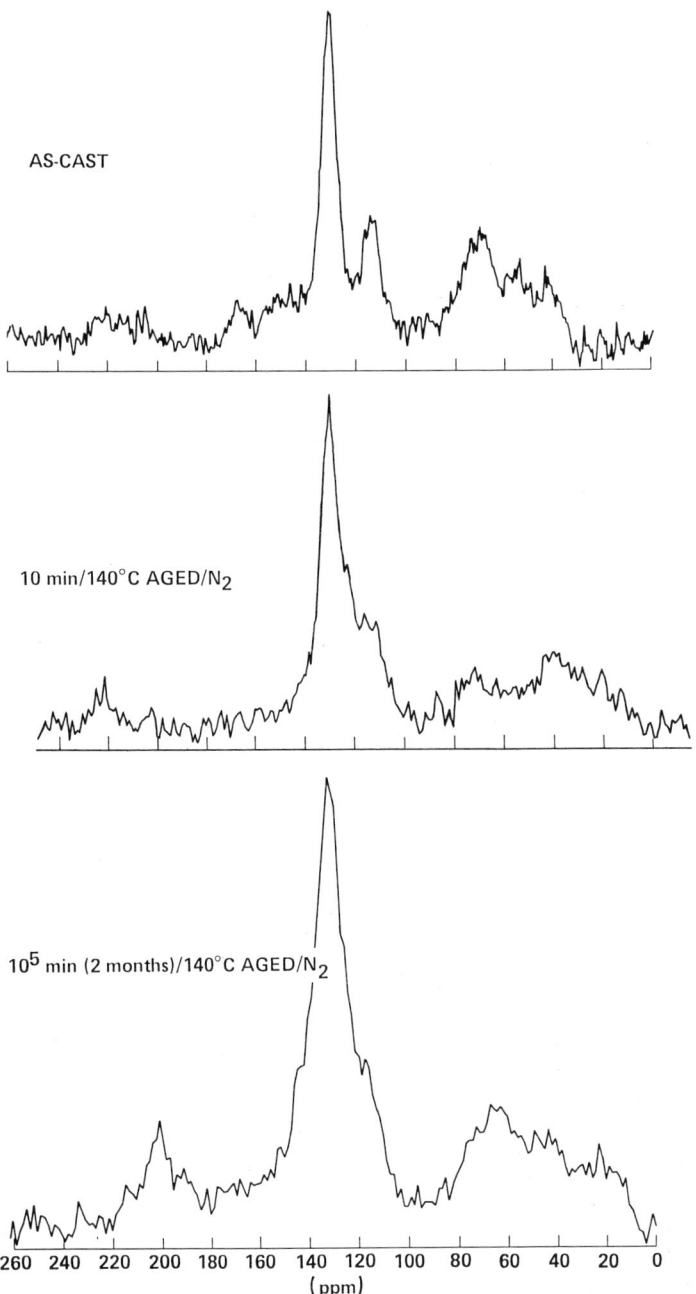

Fig. 39. Proton-decoupled CP/MAS 14.2-MHz carbon-13 NMR spectra of Fiberite 934 epoxies showing spectral changes due to postcuring and physical aging

## 3.11 Solution Carbon-13 NMR on Epoxy Components

In order to help to identify the various resonance peaks in the solid-state NMR data, it was necessary to measure the solution NMR for each of the components in the neat resin. Carbon-13 NMR spectra measured at 125 MHz of TGDDM, DDS, and DGOP dissolved in deuterated chloroform are shown in Fig. 40–44.

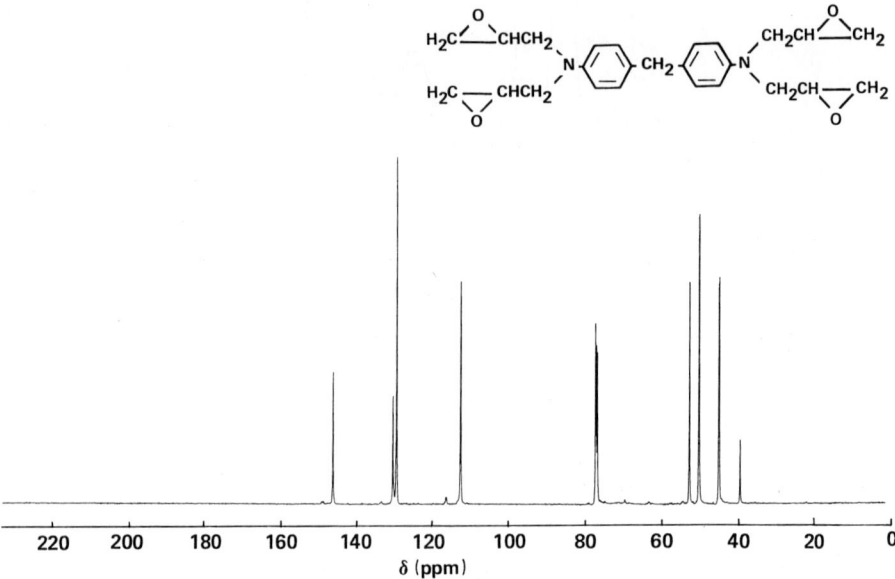

**Fig. 40.** Proton-decoupled 125 MHz carbon-13 NMR spectrum of TGDDM in deuterated chloroform solution

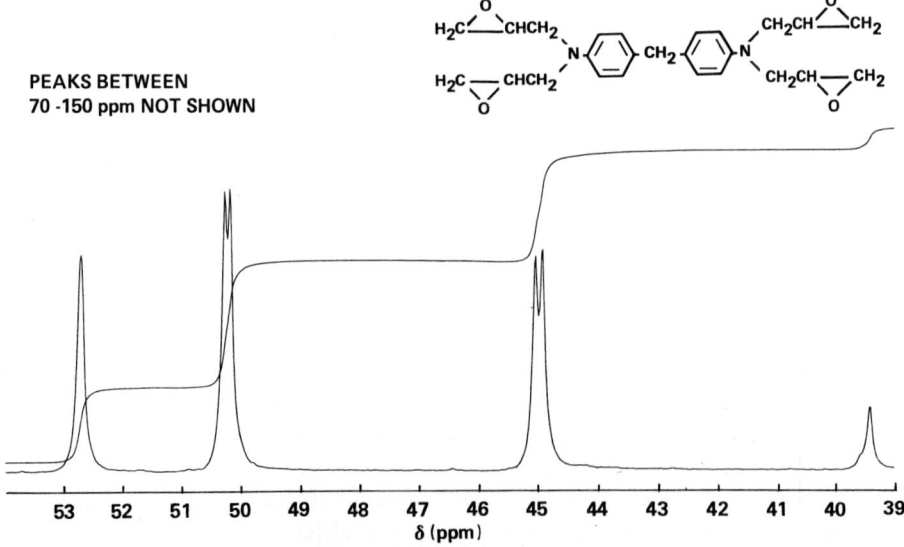

**Fig. 41.** Selected regions of the TGDDM spectrum shown in Fig. 40

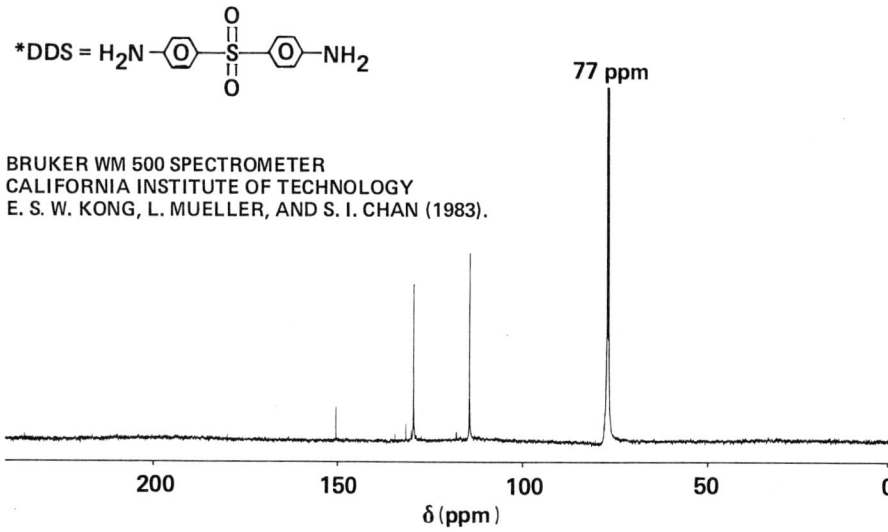

**Fig. 42.** Proton-decoupled 125 MHz carbon-13 NMR spectrum of DDS crosslinking agent in deuterated chloroform solution

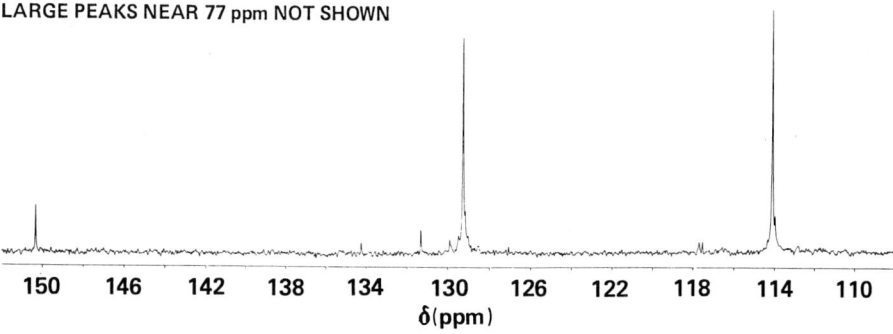

**Fig. 43.** Selected regions of the DDS spectrum shown in Fig. 42

In all three instances, the spectra show sharp resonance for the aromatics between 110 and 170 ppm and for the aliphatics between 40 and 80 ppm. The spectrum of TGDDM dissolved in deuterated chloroform contains two doubles near 45 and 50 ppm. These doublets can be attributed to conformational isomers arising from "umbrella-like" flip-flop inversions at the pyramidal-bonded nitrogen atoms [181]. In all three cases, we observed a solvent peak at 77 ppm due to the deuterated chloro-

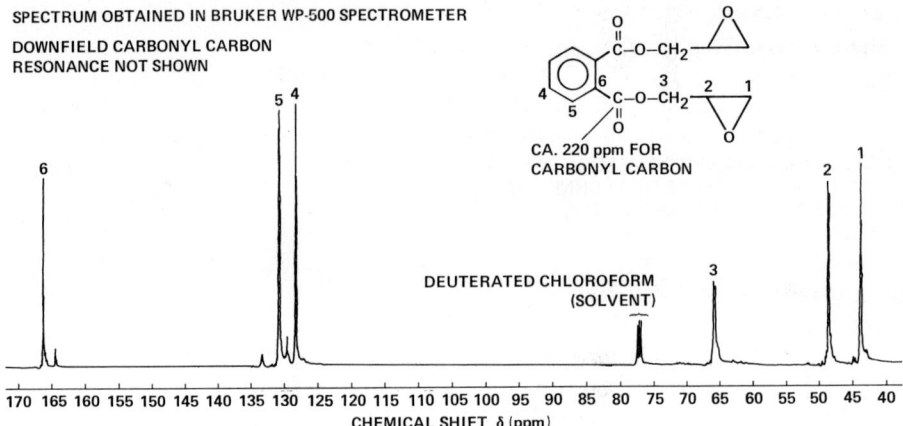

**Fig. 44.** Proton-decoupled 125 MHz carbon-13 NMR spectrum of DGOP dissolved in deuterated chloroform. Downfield carbonyl carbon resonance is not shown

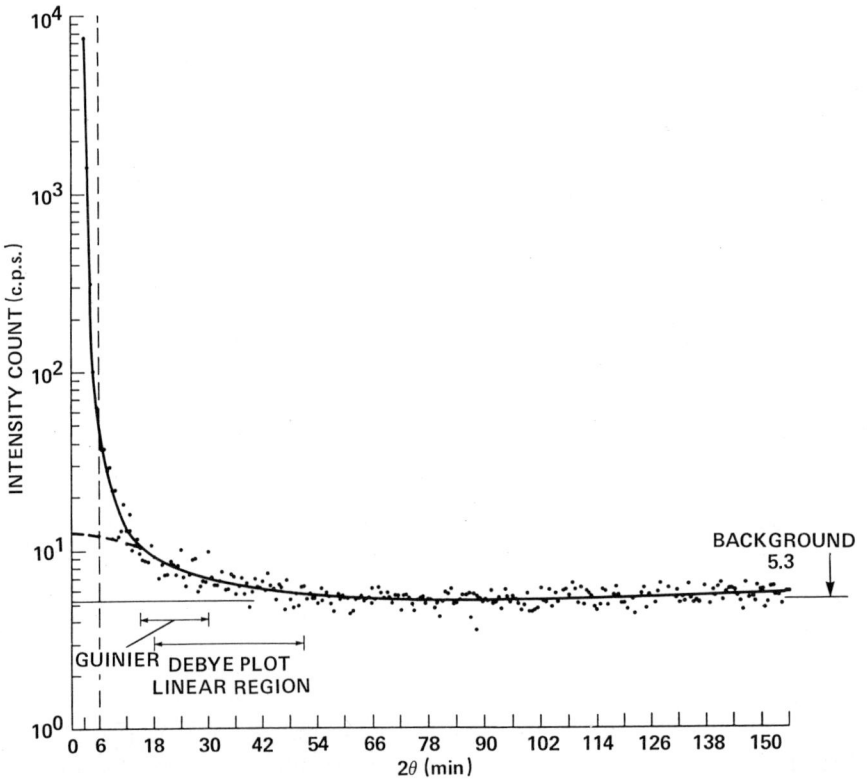

**Fig. 45.** Smeared experimental SAXS intensity from as-cast Fiberite 934 epoxy resin plotted against the scattering angle

## 3.12 SAXS Studies on Neat Resins

Figures 45 and 46 shows the logarithm of intensity count as a function of scattering angle (2 theta in minutes) for as-cast and as-postcured epoxies respectively. Specimens of the same thickness were used (0.825 mm). Guinier and Debye analysis were then carried in the regions indicated.

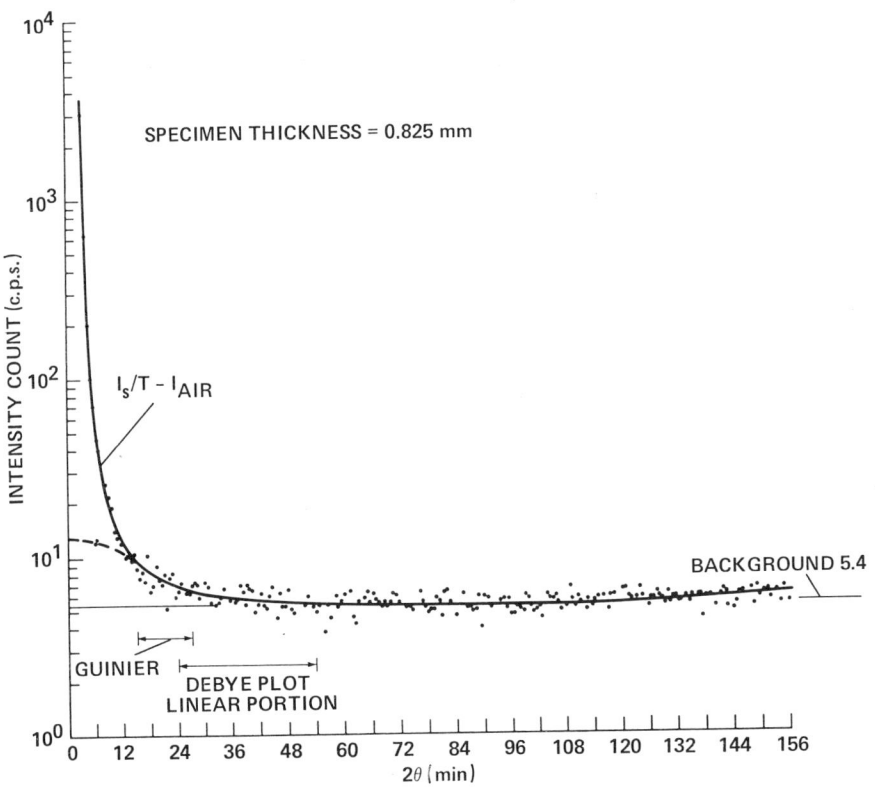

**Fig. 46.** Smeared experimental SAXS intensity of as-postcured Fiberite 934 epoxy resin plotted against the scattering angle

Desmeared SAXS data of as-cast and as-postcured resins are shown in Figs. 47 and 48. Logarithm of intensity was again plotted against the scattering angle.

Figure 49 shows the Debye plot of the SAXS data on Fiberite 934 resin. Heterogeneity of the dimension 37.0 Angstroms was observed in as-cast resin. After the postcuring schedule, the size of the heterogeneity shrinked apparently to 33.3 Ang-

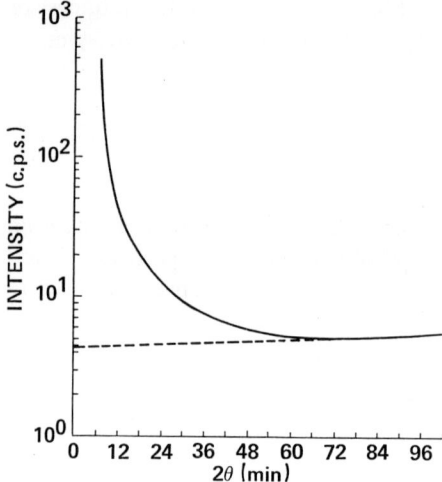

**Fig. 47.** Desmeared SAXS data of as-cast Fiberite 934 epoxy resin

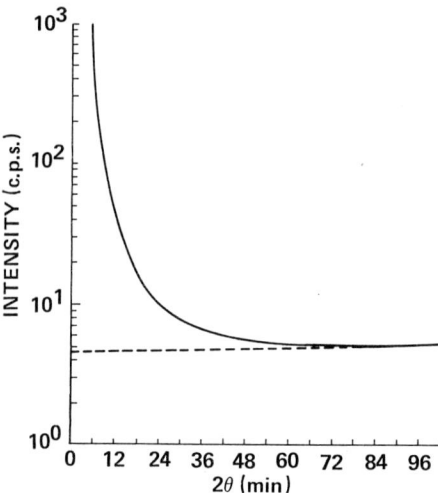

**Fig. 48.** Desmeared SAXS data of as-postcured Fiberite 934 resin

stroms. Applying the Guinier Analysis [183], however, have yielded and as-cast heterogeneity with the the dimension of 59.6 Angstroms (see Fig. 50). In this case, the resin was found to have a heterogeneity within the matrix of the size 74.2 Angstroms after postcuring (see Fig. 50).

Even though a discrepancy is seen here using the two methods of analysis in getting at the heterogeneity dimensions before and after the postcuring. One fact remains clear: there exists a heterogeneous region between the epoxy matrix of the order of 40 to 70 Angstroms. At this point, one cannot completely rule out the possibility of detecting dust particles and/or air bubbles in our SAXS experiments instead of real structural heterogeneity. However, that possibility is remote because of the repeatability of the data and the judicial use of class 10 clean room environment.

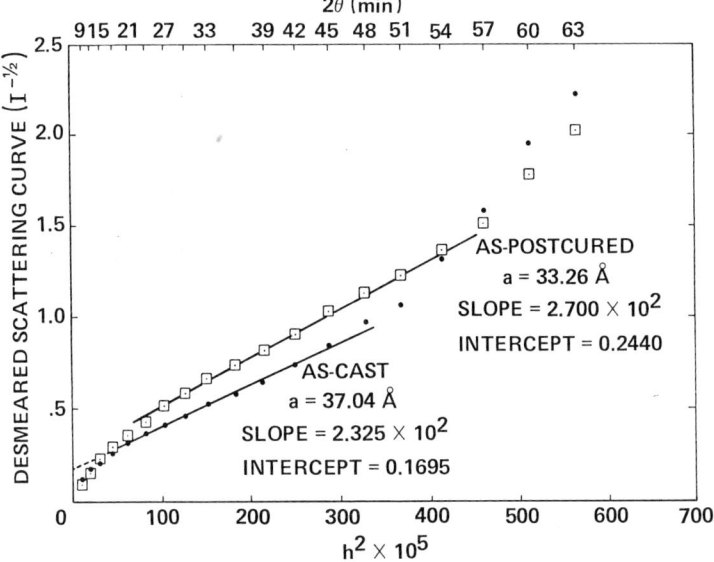

**Fig. 49.** Debye plot of SAXS data of Fiberite 934 epoxy resin

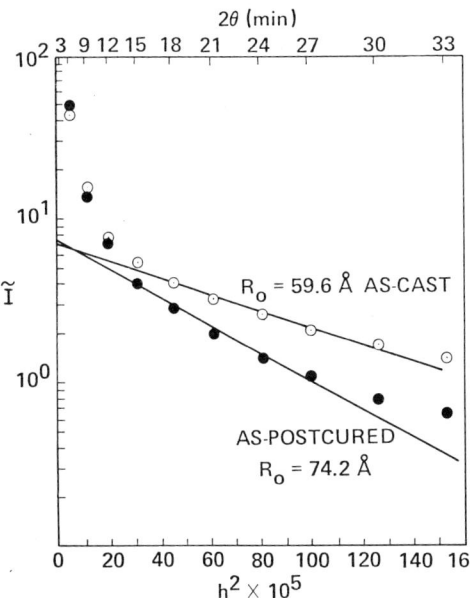

**Fig. 50.** Guinier plot SAXS data of Fiberite 934 epoxy resin

Also, utmost precaution was done in degassing most of the residual air bubbles from the polymer during casting.

Hypothetical sketches of 3 different polymer network structures are shown in Fig. 51. Model "a" indicates a network with uniform crosslink density. Model "b" indicates a network with loose ends and unconnected small network molecules. Model "c" represents a network with very non-uniform crosslink density: there exist regions of higher crosslink density embedded in a matrix of lower crosslink density. Our SAXS data have demonstrated that Fiberite 934 epoxy which contains a boron trifluoride catalyst is best represented by Model "c".

The relaxation times for polymer chains in the higher crosslink density regions are longer compared to those in lower crosslink density regions. The physical aging kinetics are therefore different depending on the crosslink density. With this idea in mind, experiments are currently performed using SAXS and positron annihilation [186, 187] techniques to correlate the effects of crosslink density and free volume distributions on physical aging kinetics [188, 189].

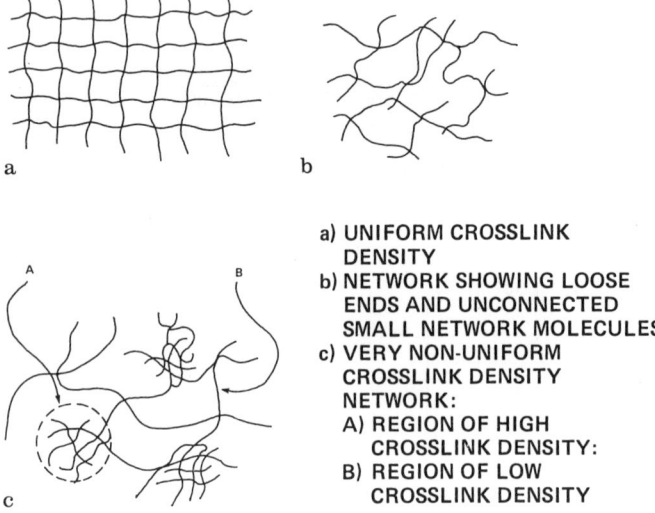

a) UNIFORM CROSSLINK DENSITY
b) NETWORK SHOWING LOOSE ENDS AND UNCONNECTED SMALL NETWORK MOLECULES
c) VERY NON-UNIFORM CROSSLINK DENSITY NETWORK:
　A) REGION OF HIGH CROSSLINK DENSITY:
　B) REGION OF LOW CROSSLINK DENSITY

**Fig. 51.** Hypothetical two dimensional sketches of three different polymer network structures

# 4 Conclusions

For the first time, it is possible to characterize the molecular aggregation during physical aging in network epoxies by spectroscopic and other techniques. Investigations were made in a network with structural heterogeneity as indicated by SAXS studies. Most of the observations can be rationalized using a qualitative concept of free volume. The important aspects of the findings are summarized: as physical aging proceeds, the following properties of the polymer changes.

- Damping decreases
- Ultimate mechanical properties decrease
- Stress relaxation rates decrease

- Moisture sorption decreases
- Moisture diffusivity decreases
- Density increases
- Dynamic modulus increases
- Hardness increases
- Moisture/epoxy swelling interaction increases
- Glassy-state expansivity decreases
- Rubbery-state expansivity increases

The thermoreversible nature of physical aging has been amply demonstrated.

*Acknowledgements*: This research has been supported by a grant from NASA (NCC 2-103). The use of the Southern California Regional NMR Facility is gratefully acknowledged. This NMR Facility is supported by NSF Grant No. CHE 79-16324. The author is grateful to Professor S. I. Chan for his valuable comments and constructive reviews. The use of the SAXS facility in Kyoto University was possible through the courtesy of Professor H. Kawai and Dr. T. Hashimoto. The author would also like to thank Dr. L. Mueller for measuring the solid state NMR data and Mr. M. Adamson for performing some of the diffusion experiments. Finally, the technical support from Ms. Susanna Lee is gratefully acknowledged.

# 5 References

1. Simon, F.: Z. Anorg. Allgem. Chem. *203*, 219 (1931)
2. Qayyum, M. M., White, J. R.: J. Polym. Sci.; Polym. Lett. Ed. *21*, 31 (1983)
3. Frosini, V., Levita, G., Marchetti, A.: J. Calorim. Analys. Therm. *9A*, B5–35 (1978)
4. Merrall, G. T., Meeks, A. C.: J. Appl. Polym. Sci. *16*, 3389 (1972)
5. Su, C. J.: Polym. Plast. Technol. Eng. *4*, 221 (1975)
6. Wyzgoski, M. C.: Polym. Eng. Sci. *16*, 265 (1976)
7. Uchidoi, D., Adachi, K., Ishida, Y.: Polym. J. *10*, 161 (1978)
8. Moynihan, C. T.: Ann. New York Acad. Sci. *279*, 15 (1976)
9. Struik, L. C. E.: La Chimica E L'Industria *58*, 549 (1976)
10. Struik, L. C. E.: Proceedings of the 7th International Congress on Rheology, Chalmers University of Technology, Gothenburg, Sweden, Pages 23–27, August (1976)
11. Struik, L. C. E.: Polym. Eng. Sci. *17*, 165 (1977)
12. Tant, M. R., Wilkes, G. L.: Polym. Eng. Sci. *21*, 874 (1981)
13. Booij, H. C., Minkhorst, J. H. K.: Polym. Eng. Sci. *19*, 579 (1979)
14. Struik, L. C. E.: Polym. Eng. Sci. *18*, 799 (1978)
15. Chow, T. S., Prest Jr., W. M.: J. Appl. Phys., *53*, 6568 (1982)
16. Chai, C. K., McCrum, N. G.: Polymer *21*, 706 (1980)
17. Kaestner, S., Dittmer, M.: Koll. Z. Z. Polym. *204*, 74 (1965)
18. Wyzgoski, M. G.: J. Appl. Polym. Sci. *25*, 1443 (1980)
19. Kovacs, A. J., Stratton, R. A., Ferry, J. D.: J. Phys. Chem. *67*, 152 (1963)
20. Frank, W., Goddar, H., Stuart, H. A.: J. Polym. Sci., Polym. Letters, Part B *5*, 711 (1967)
21. Wyzgoski, M. G.: J. Appl. Polym. Sci. *25*, 1455 (1980)
22. Bauwens-Crowet, C., Bauwens, J. C.: Polymer *23*, 1599 (1982)
23. LeGrand, D. G.: J. Appl. Polym. Sci. *13*, 2129 (1969)
24. Ender, D. H.: J. Macromol. Sci.-Phys. *B4*, 635 (1970)

25. Chang, T. D., Carr, S. H., Brittain, J. O.: Bull. Amer. Phys. Soc. 27, no. 3, 290 (1982)
26. Chang, T. D., Carr, S. G., Brittain, J. O.: Polym. Eng. Sci. 22, 1205 (1982)
27. Chang, T. D., Carr, S. H., Brittain, J. O.: Polym. Eng. Sci. 22, 1213 (1982)
28. Chang, T. D., Brittain, J. O.: Polym. Eng. Sci. 22, 1221 (1982)
29. Chang, T. D., Brittain, J. O.: Polym. Eng. Sci. 22, 1228 (1982)
30. Aref-Azar, A., Hay, J. N.: Polymer 23, 1129 (1982)
31. Levita, G., Smith, T. L.: Polym. Eng. Sci. 21, 936 (1981)
32. Kaiser, J.: Makromol. Chem. 180, 573 (1979)
33. Yokouchi, M., Kobayashi, Y.: J. Appl. Polym. Sci. 26, 431 (1981)
34. Cama, F. J., Morbitzer, L.: Soc. Plast. Eng. Tech Papers 28, 96 (1982)
35. Chan, A. H., Paul, D. R.: J. Appl. Polym. Sci. 24, 1539 (1979)
36. Chan, A. H., Paul, D. R.: Polym. Eng. Sci. 20, 87 (1980)
37. Morgan, R. J., O'Neal, J. E.: J. Polym. Sci., Polym. Phys. Ed. 14, 1053 (1976)
38. Chan, A. H., Paul, D. R.: J. Appl. Polym. Sci. 25, 971 (1980)
39. Neki, K., Geil, P. H.: in "The Solid State of Polymers". edited by P. H. Geil, E. Baer, and Y. Wada, pp. 295–341, Dekker, New York (1974)
40. Golden, J. H., Hammant, B. L., Hazell, E. A.: J. Appl. Polym. Sci. 11, 1571 (1967)
41. Mininni, R. M., Moore, R. S., Flick, J. R., Petrie, S. E. B.: J. Macromol. Sci., Phys. B8, 343 (1973)
42. Mininni, R. M., Moore, R. S., Flick, J. R., Petrie, S. E. B.: in The Solid State of Polymers., ed. Geil, P. H., Baer, E., and Wada, Y., pp. 343–359, New York: Dekker 1974
43. Peilstoecker, G.: Kunststoffe 51, 509 (1961)
44. Peilstoecker, G.: Brit. Plastics 35, 365 (1962)
45. Ito, E., Yamamoto, K., Kobayashi, Y., Hatakeyama, T.: Polymer 19, 39 (1978)
46. Rizzo, G., Titomanlio, G.: Polym. Bull. 1, 743 (1979)
47. Meyer, H. H., Mangin, P. M. F., Ferry, J. D.: J. Polym. Sci., Part A 3, 1785 (1965)
48. Chai, C. K., McCrum, N. G.: Polymer 25, 291 (1984)
49. Struik, L. C. E.: Plast. Rubber Process. Appl. 2, 41 (1982)
50. Adachi, K., Kotaka, T.: Polym. J. 14, 959 (1982)
51. Hozumi, S.: Polym. J. 1, 632 (1970)
52. Hozumi, S., Wakabayashi, T., Sujihara, K.: Polym. J. 2, 756 (1971)
53. Sung, C. S. P., Lamarre, L., Chung, K. H.: Macromolecules 14, 1839 (1981)
54. Joss, B. L., Bretzlaff, R. S., Wool, R. P.: J. Appl. Phys. 54, 5515 (1983)
55. Tant, M. R., Wilkes, G. L.: Polym. Eng. Sci. 21, 325 (1981)
56. Matsuoka, S., Bair, H. F.: J. Appl. Phys. 48, 4058 (1977)
57. Matsuoka, S., Bair, H. E., Bearder, S. H., Kern, H. E., Ryan, J. T.: Polym. Engin. Sci. 18, 1073 (1978)
58. Struik, L. C. E.: Physical Aging in Amorphous Polymers and Other Materials. Amsterdam: Elsevier 1978
59. Struik, L. C. E.: Rheol. Acta 5, 303 (1966)
60. Bree, H. W., Heijboer, J., Struik, L. C. E., Tak, A. G. M.: J. Polym. Sci., Polym. Phys. Ed. 12, 1857 (1974)
61. Struik, L. C. E.: Ann. New York Acad. Sci. 279, 78 (1976)
62. Struik, L. C. E.: Polymer 21, 962 (1980)
63. Struik, L. C. E.: Europhys. Conf. Abstr. 4A, 135 (1980)
64. Ophir, Z. H., Emerson, J. A., Wilkes, G. L.: J. Appl. Phys. 49, 5032 (1978)
65. Kovacs, A. J.: J. Polym. Sci. 30, 131 (1958)
66. Kovacs, A. J.: Adv. Polym. Sci. 3, 394 (1963)
67. Kovacs, A. J., Stratton, R. A., Ferry, J. D.: J. Phys. Chem. 67, 152 (1963)
68. Kovacs, A. J., Hutchinson, J. M., Aklonis, J. J.: in "The Structure of Non-crystalline Materials", ed. Gaskell, P. H., pp. 153–163, London: Taylor and Francis 1977
69. Aklonis, J. J., Kovacs, A. J.: in Contemporary Topics in Polymer Science, Volume 3. ed. Shen, M., pp. 267–295, New York: Plenum 1979
70. Kovacs, A. J., Aklonis, J. J., Hutchinson, J. M., Ramos, A. R.: J. Polym. Sci., Polym. Phys. Ed. 17, 1097 (1979)
71. Petrie, S. E. B.: J. Polym. Sci., A-2, 10, 1255 (1972)

72. Petrie, S. E. B.: in "Polymeric Materials: Relationships between Structure and Mechanical Behavior". ed. Baer, E. and Radcliffe, S. V., pp. 55–118, American Society for Metals, Ohio: Metals Park, 1975
73. Marshall, A. S., Petrie, S. E. B.: J. Appl. Phys. *46*, 4223 (1975)
74. Petrie, S. E. B.: J. Macromol. Sci. Phys. *B12*, 225 (1976)
75. Myers, F. A., Cama, F. C., Sternstein, S. S.: Ann. New York Acad. Sci. *279*, 94 (1976)
76. Sternstein, S. S., Ho, T. C.: J. Appl. Phys. *43*, 4370 (1972)
77. Sternstein, S. S.: in "Treatise on Materials Science and Technology, Volume 10, Part B". ed. Schultz, J. M., pp. 541–598, New York: Academic Press 1977
78. Kibler, K. G.: Time-dependent Environmental Behavior of Graphite/Epoxy Composites. U.S. Air Force Wright Aeronautical Laboratories Report AFWAL-TR-80-4052 (1980)
79. Stuart, H. A.: in Alterung und Korrosion von Kunststoffen. Korrosion 20, pp. 87–94, Weinheim: Verlag Chemie 1967
80. Hedvig, P.: "Dielectric Spectroscopy of Polymers." New York: Wiley 1977
81. Davies, M. (ed.): Dielectric and Related Molecular Processes: Volume 3. London: The Chemical Society 1977
82. Tareev, B.: "Physics of Dielectric Materials." Moscow: Mir Publishers 1975
83. Hill, N., Vaughan, W. E., Price, A. H., Davies, M.: Dielectric Properties of Molecular Behaviour. London: Van Nostrand Reinhold 1969
84. Tobolsky, A. V.: Properties and Structure of Polymers. New York: Wiley (1980)
85. Aklonis, J. J., MacKnight, W. J.: Introduction to Polymer Viscoelasticity. Second edition, New York: Wiley-Interscience 1983
86. Bueche, F.: Physical Properties of Polymers. New York: Interscience 1962
87. Hiemenz, P. C.: Polymer Chemistry: The Basic Concepts. pp. 248–256. New York: Marcel Dekker 1984
88. Curro, J. G., Lagasse, R. R., Simha, R.: J. Appl. Phys. *52*, 5892 (1981)
89. Meissner, B.: J. Polym. Sci., Polym. Lett. Ed. *19*, 137 (1981)
90. Haward, R. N.: Rev. Macromol. Chem. *5*, Part-2, 45 (1970)
91. Haward, R. N. (ed.): The Physics of Glassy Polymers. New York: Wiley 1973
92. Eyring, H.: J. Chem. Phys. *4*, 283 (1936)
93. Williams, W. L., Landel, R. F., Ferry, J. D.: J. Amer. Chem. Soc. *77*, 3701 (1955)
94. Rusch, K. C., Beck, R. H.: Plastic Deformation of Polymers. ed. Peterlin, A., pp. 161–173, New York: Marcel Dekker 1971
95. Rusch, K. C., Beck, R. H.: J. Macromol. Sci., Phys. *B3*, 365 (1969)
96. Lipatov, Y.: Adv. Polym. Sci. *26*, 63 (1978)
97. Doolittle, A. K.: J. Appl. Polym. Sci. *24*, 1329 (1979)
98. Vrentas, J. S., Duda, J. L.: J. Polym. Sci., Polym. Phys. Ed. *15*, 403 (1977)
99. Yoshida, H., Kobayashi, Y.: Polym. J. *14*, 925 (1982)
100. Bendler, J. T., Nagi, K. L.: Macromolecules *17*, 1174 (1984)
101. Yilmazer, U., Farris, R. J.: J. Appl. Polym. Sci. *28*, 3269 (1983)
102. Chow, T. S.: J. Chem. Phys. *79*, 4602 (1983)
103. Lamarre, L., Sung, C. S. P.: Macromolecules *16*, 1729 (1983)
104. Hodge, I. M.: Macromolecules *16*, 898 (1983)
105. Hodge, I. M., Huvard, G. S.: Macromolecules *16*, 371 (1983)
106. Prest, W. M., O'Reilly, J. M., Roberts, F. J., Mosher, R. A.: Polym. Eng. Sci. *21*, 1181 (1981)
107. Prest, W. M., Roberts, F. J.: Ann. New York Acad. Sci. *371*, 67 (1981)
108. Chow, T. S., Prest, W. M.: J. Appl. Phys. *53*, 6568 (1982)
109. Yoshida, H., Nakamura, K., Kobayashi, Y.: Polym. J. *14*, 855 (1982)
110. Kong, E. S. W., Wilkes, G. L., McGrath, J. E., Banthia, A. K., Mohajer, Y., Tant, M. R.: Polym. Eng. Sci. *21*, 943 (1981)
111. Kong, E. S. W.: J. Appl. Phys. *52*, 5921 (1981)
112. Kong, E. S. W.: Composites Technology Rev. *4*, 97 (1982)
113. Kong, E. S. W., Adamson, M. J.: Polymer Commun. *24*, 171 (1983)
114. Kong, E. S. W.: in "Epoxy Resin Chemistry II." ed. Bauer, R. S., Amer. Chem. Soc. Symp. Series no. 221, Chapter 9, pp. 171–191, Washington, D.C.: Amer. Chem. Soc., 1983
115. Kong, E. S. W.: in Contemporary Topics in Polymer Science, Volume 4. ed. Bailey, W. J., and Tsuruta, T., pp. 629–646, New York: Plenum Press 1984

116. Kong, E. S. W.: in Characterization of Highly Crosslinked Polymers. ed. Labana, S. S., and Dickie, R. A., Amer. Chem. Soc. Symp. Series no. 243, Chapter 9, pp. 125–164, Washington, D.C.: Amer. Chem. Soc. 1984
117. Kong, E. S. W., Adamson, M., Mueller, L.: Composites Technol. Rev., 6, 170 (1984)
118. Simha, R., Curro, J. G., Robertson, R. E.: Polym. Eng. Sci., 24, 1071 (1984)
119. Chow, T. S.: Polym. Eng. Sci., 24, 1079 (1984)
120. Hutchinson, J. M., Kovacs, A. J.: Polym. Eng. Sci., 24, 1087 (1984)
121. Rendell, R. W., Lee, T. K., Ngai, K. L.: Polym. Eng. Sci., 24, 1104 (1984)
122. Plazek, D. J., Ngai, K. L., Rendell, R. W.: Polym. Eng. Sci. 24, 1111 (1984)
123. Moynihan, C. T., Bruce, A. J., Gavin, D. L., Loehr, S. R., Opalka, S. M., Drexhage, M. G.: Polym. Eng. Sci., 24, 1117 (1984)
124. Berens, A. R., Hodge, I. M.: Polym. Eng. Sci., 24, 1130 (1984)
125. Joss, B. L., Bretzlaff, R. S., Wool, R. P.: Polym. Eng. Sci., 24, 1130 (1984)
126. McKenna, G. B., Kovacs, A. J.: Polym. Eng. Sci., 24, 1138 (1984)
127. Rubeck, R. A., Bales, S. E., Lee, H. D.: Polym. Eng. Sci., 24, 1142 (1984)
128. Abkowitz, M.: Polym. Eng. Sci., 24, 1149 (1984)
129. Fukushima, E., Roeder, S. B. W.: Experimental Pulse NMR. Reading, Massachusetts: Addison-Wesley 1981
130. Browning, C. E.: Polym. Eng. Sci., 18, 16 (1978)
131. Morgan, R. J., O'Neal, J. E., Fanter, D. L.: J. Mater. Sci., 15, 751 (1980)
132. Wright, W. W.: Composites, 12, 201 (1981)
133. Ellis, T. S., Karasz, F. E.: Polymer, 25, 644 (1984)
134. Jelinski, L. W., Dumais, J. J., Stark, R. E., Ellis, T. S., Karasz, F. E.: Macromolecules, 16, 1019 (1983)
135. May, C. A., Fritzen, J. S., Whearty, D. K.: Exploratory Development of Chemical Quality Assurance and Composition of Epoxy Formulations. Lockheed Missiles and Space Company, Air Force Technical Report: AFML-TR-76-112 (1976)
136. Hadad, D. K., Fritzen, J. S., May, C. A.: Exploratory Development of Chemical Quality Assurance and Composition of Epoxy Formulations. Lockheed Missiles and Space Company, Air Force Technical Report: AFML-TR-77-217 (1977)
137. Adamson, M. J.: J. Mater. Sci., 15, 1736 (1980)
138. Hendricks, R. W., Schmidt, P. W.: Acta Phys. Austriaca, 37, 20 (1973)
139. Schmidt, P. W.: Acta Crystallogr., 19, 938 (1965)
140. Kong, E. S. W., Wilkes, G. L.: Internal Report written on a dental restorative resin supplied by Johnson and Johnson Company, New Brunswick, New Jersey (1979)
141. Agarwal, B. D., Broutman, L. J.: Analysis and Performance of Fiber Composites. New York: Wiley 1980
142. Heijboer, J.: Annals New York Acad. Sci., 279, 104 (1976)
143. Takahama, T., Geil, P. H.: J. Polym. Sci., Phys. Ed., 20, 1979 (1982)
144. Bailey, R. T., Noeth, A. M., Pethrick, R. A.: Molecular Motion in High Polymers. Oxford: Oxford University Press, Page 287 (1981)
145. Kaelble, D. H.: J. Appl. Polym. Sci., 9, 213 (1965)
146. May, C. A., Weir, F. E.: Soc. Plastics Eng. Transactions, 7, 207 (1962)
147. Browning, C. E.: Polym. Eng. Sci., 18, 16 (1978)
148. Murayama, T., Bell, J. P.: J. Polym. Sci. A-2, 8, 437 (1970)
149. Kalfoglou, N. K., Williams, H. L.: J. Appl. Polym. Sci., 17, 1377 (1973)
150. Cook, W. D., Delatycki, O.: J. Polym. Sci., Polym. Phys. Ed., 13, 1049 (1975)
151. Murayama, T.: Dynamic Mechanical Analysis of Polymeric Material. Amsterdam: Elsevier 1978
152. Wyzgoski, M. G.: J. Appl. Polym. Sci., 25, 1443 (1980)
153. Willbourn, A. H.: Trans. Faraday Soc., 54, 717 (1958)
154. Kenyon, A. S., Nielsen, L. E.: J. Macromol. Sci., Chem. Ed., A3, 275 (1969)
155. Chang, T. D., Carr, S. H., Brittain, J. O.: Polym. Eng. Sci., 22, 1205 (1982)
156. Nielsen, L. E.: Soc. Plastics Engin. Trans. 16, 525 (1960)
157. Read, B. E., Dean, G. D.: The Determination of Dynamic Properties of Polymers and Composites. Bristol: Adam Hilger 1978

158. Von Kuzenko, M., Browning, C. E.: Amer. Chem. Soc. Preprints Org. Coat. Plastics Chem., 40, 694 (1979)
159. Keenan, J., Seferis, J. C., Quinlivan, J. T.: J. Appl. Polym. Sci., 24, 2375 (1979)
160. Hata, N., Yamaguchi, R., Kumanotani, J.: J. Appl. Polym. Sci., 17, 2173 (1973)
161. Wetton, R. E.: Anal. Proc., Anal. Div. Royal Soc., Chem., Page 416, October (1981)
162. Bank, L., Ellis, B.: Polym. Bull., 1, 377 (1979)
163. Schatzki, T. F.: J. Polym. Sci., 57, 496 (1962)
164. Wunderlich, B.: J. Chem. Phys., 37, 2429 (1962)
165. Boyer, R. F.: Rubber Rev., 34, 1303 (1963)
166. Pogany, G. A.: Polymer, 11, 66 (1970)
167. Roberts, G. E., White, E. F. T.: The Physics of Glassy Polymers. ed. Haward, R. N., New York: Wiley, Page 153, 1973
168. Meier, D. J., (ed.): Molecular Basis of transitions and Relaxations. Midland Macromolecular Monographs, Vol. 4, New York: Gordon and Breach Science Publishers 1978
169. Morgan, R. J., Happe, J. A., Mones, E. T.: Proceedings 28th National SAMPE Symposium Page 596 (1983)
170. Joesten, M. D., Schaad, L. J.: Hydrogen Bonding. New York: Dekker 1974
171. Pimentel, G. C., McClellan, A. L.: The Hydrogen Bond. San Francisco: Freeman 1960
172. Moynihan, C. T., Macedo, P. B., Montrose, C. J., Gupta, P. K., DeBolt, M. A., Dill, J. F., Dom, B. E., Drake, P. W., Easteal, A. J., Elterman, P. B., Moeller, R. P., Sasabe, H., Wilder, J. A.: Ann. New York Acad. Sci., 279, 15 (1976)
173. Qi, Z.: private communications (1983)
174. Ophir, Z.: Structure-Property Relationships in Solid Polymers: I — Segmented Polyurethanes and II — Epoxy Thermosets. Ph. D. Thesis, Princeton University (1979)
175. Aherne, J. P., Enns, J. B., Doyle, M. J., Gillham, J. K.: Amer. Chem. Soc., Org. Coat. Appl. Polym. Sci. Proc., 46, 574 (1982)
176. Enns, J. B., Gillham, J. K.: J. Appl. Polym. Sci., 28, 2831 (1983)
177. Jelinski, L., et al.: Macromolecules, in press (1985)
178. Garcia-Fierro, J. L., Aleman, J. V.: Macromolecules, 15, 1145 (1982)
179. Shen, C. H., Springer, G. S.: J. Composite Mater., 10, 2 (1976)
180. Levy, G. C., Nelson, G. L.: Carbon-13 Nuclear Magnetic Resonance for Organic Chemists. New York: Wiley (1972)
181. Morrison, R. T., Boyd, R. N.: Organic Chemistry. 4th edition, Boston: Allyn and Bacon 1983
182. Kakudo, M., Kasai, N.: X-Ray Diffraction by Polymers. Amsterdam: Elsevier 1972
183. Guinier, A.: X-Ray Diffraction in Crystals, Imperfect Crystals, and Amorphous Bodies. San Francisco: Freeman 1963
184. Glatter, O., Kratky, O. (eds.): Small Angle X-Ray Scattering. New York: Academic Press 1982
185. Alexander, L. E.: X-ray Diffraction in Polymer Science. New York: Wiley 1969
186. Djermouni, B., Ache, H. J.: Macromolecules, 13, 168 (1980)
187. Ache, H. J. (ed.): Positronium and Muonium Chemistry. Advances in Chemistry Series No. 175, Washington, D.C.: American Chemical Society 1979
188. Robertson, R. E., Simha, R., Curro, J. G.: Macromolecules, 18, 2239 (1985)
189. Fredrickson, G. H., Helfand, E.: Macromolecules, 18, 2201 (1985)

Editor: K. Dušek
Received August 13, 1985

# Curing Mechanisms and Mechanical Properties of Cured Epoxy Resins

Takashi Kamon
Kyoto Municipal Research Institute of Industry

Hitoshi Furukawa
Sanyo Giken (Sanyo Technical Research), Limited
1-2 Minamida-cho, Nishikujo, Minami-ku, Kyoto 601/Japan

*Epoxy resins can be cured with various kinds of hardeners, which results in many types of epoxy resins with different structures. Here, the relationships between the structures and mechanical properties, particularly dynamic mechanical properties of those resins are reviewed. The structures hardly affect the mechanical properties in the glassy region. The mechanical properties drop rapidly at temperatures higher than the glass transition temperature ($T_g$). The relationships between the resin structures and $T_g$ are discussed. The factors affecting $T_g$ such as the type of networks, the molecular structure of the initial starting raw materials, and the density of crosslinking (number of functional groups), are considered here.*

*Furthermore, mechanical properties of some cured resins having structures which cannot be presumed from initial resins are also discussed. Among the mechanical properties, impact strength is related to $\beta$ dispersion at around $-50\,°C$ rather than to dispersion ($T_g$).*

1 Introduction . . . . . . . . . . . . . . . . . . . 174

2 The Curing Mechanism of Epoxy Resins and the Structure of Cured Resins  174

3 Types of Curing Agents and the Dynamic Mechanical Properties of Cured Resins  177
   3.1 Dynamic Mechanical Properties of Cured Resins with Different Curing Mechanisms . . . . . . . . . . . . . . . . . . . 179
   3.2 Structures and Dynamic Mechanical Properties of Epoxy Resins Cured with Polyamines . . . . . . . . . . . . . . . . . . . 180
   3.3 Dynamic Mechanical Properties of Resins Cured with Carboxylic Hydrazides . . . . . . . . . . . . . . . . . . . 183
   3.4 Dynamic Mechanical Properties of Epoxy Resins Cured with Carboxylic Anhydride . . . . . . . . . . . . . . . . . . . 186
   3.5 Dynamic Mechanical Properties of Mercaptan-cured Resin . . . . . . 188

4 Dynamic Mechanical Properties of Resins Cured with Different Hardeners . . 190
   4.1 Dicyandiamide . . . . . . . . . . . . . . . . . . . 190
   4.2 Resins Cured with Isocyanates . . . . . . . . . . . . . . . . . . . 192
   4.3 Other Examples . . . . . . . . . . . . . . . . . . . 193

5 Mechanical Properties of Cured Epoxy Resins in the Glassy State . . . . . 194

6 Conclusions . . . . . . . . . . . . . . . . . . . 199

7 Appendix . . . . . . . . . . . . . . . . . . . 202

8 References . . . . . . . . . . . . . . . . . . . 202

# 1 Introduction

Epoxy resins are superior in heat resistance, adhesion, corrosion resistance and also mechanical properties among thermosetting resins and are widely used for coatings, adhesives, electric insulating materials and matrices for FRP in areas such as aircrafts, electronics, electric power, and building and civil engineering.

However, epoxy resins now in use are not of the same structures as thermoplastic resins.

The factors determining the structure of cured resins and affecting the physical and mechanical properties are as follows:
1. Curing mechanism: kind of functional groups of hardeners.
2. Number of functional groups in resins and hardeners: density of crosslinking.
3. Molecular structure of bridges between functional groups in resins and hardeners.
4. Molar ratio of resin and hardener: density of crosslinking.
5. Degree of curing, or curing conditions.

Most epoxy resin prepolymers now in industrial use are diglycidyl ethers having two epoxy groups per molecule; a few tens of other epoxy resin prepolymers are now also available on the market. On the other hand, more than one hundred hardeners are known. Therefore, the number of combinations of resins and hardeners is very high and, accordingly, it is hardly possible to describe all mechanical properties of the cured resins with many different structures — hardness, tensile strength, elongation at break, impact strength, etc. within a wide range of temperatures.

A number of papers deals with the dynamic mechanical properties of cured epoxy resins [1-5] and with the effect of the structure of basic resins and the degree of crosslinking [6-8].

In this review, the effect of the cured epoxy resins on the dynamic mechanical properties of epoxy resins is discussed.

# 2 The Curing Mechanism of Epoxy Resins and the Structure of Cured Resins

The mechanical properties and dynamic mechanical properties of the cured epoxy resins are governed by their structures.

The curing mechanism of an epoxy resin or the type of functional group of a hardener is the most essential factor determining the structure of the cured resin. The well-known hardeners are polyamines, acid anhydrides, and polymerization catalysts.

Among these hardeners, amines are the most versatile ones at room temperature as well as elevated curing temperature. The curing mechanisms with amines and the structures of the amine cured epoxy resins have been most sufficiently studied, and the systems of epoxy resins with amine hardeners are most extensively used in the practical industrial fields.

# Curing Mechanisms and Mechanical Properties of Cured Epoxy Resins

The curing mechanism of a resin with a primary amine is as follows [9-12]:

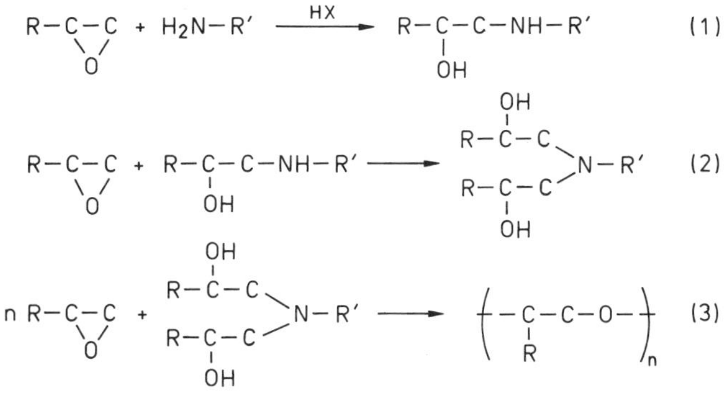

$$R-\underset{O}{C-C} + H_2N-R' \xrightarrow{HX} R-C-C-NH-R' \quad (1)$$
$$\phantom{XXXXXXXXXXXXXX} \underset{OH}{|}$$

$$R-\underset{O}{C-C} + R-\underset{OH}{C}-C-NH-R' \longrightarrow \begin{array}{c} R-\underset{OH}{C}-C \\ \phantom{X}\diagdown \\ R-\underset{OH}{C}-C \end{array} N-R' \quad (2)$$

$$n\ R-\underset{O}{C-C} + \begin{array}{c} R-\underset{OH}{C}-C \\ \phantom{X}\diagdown \\ R-\underset{OH}{C}-C \end{array} N-R' \longrightarrow \left(-C-\underset{R}{C}-O-\right)_n \quad (3)$$

As shown in the above reaction scheme, the epoxy groups are successively opened by amine active hydrogens [Eqs. (1) and (2)].

In this reaction, the presence of active hydrogen compounds (HX), such as water and alcohols as impurities, is required, and alcohols produced by Eqs. (1) and (2) accelerate the curing reaction (cf., e.g., Smith [9] and Kakurai [11]). Different mechanisms have been proposed by Smith [9], Tanaka [11], and Bell [12], but they have not yet been confirmed.

It has been shown, however, that curing proceeds as shown in Eqs. (1) and (2), and the structures shown in Fig. 1a are finally formed. It is generally accepted that Eq. (3) hardly occurs in case of the stoichiometrically equivalent system or with an excess of amine [13].

In addition of polyamines, acid anhydride hardeners are used but the structure of cured resins is less understood.

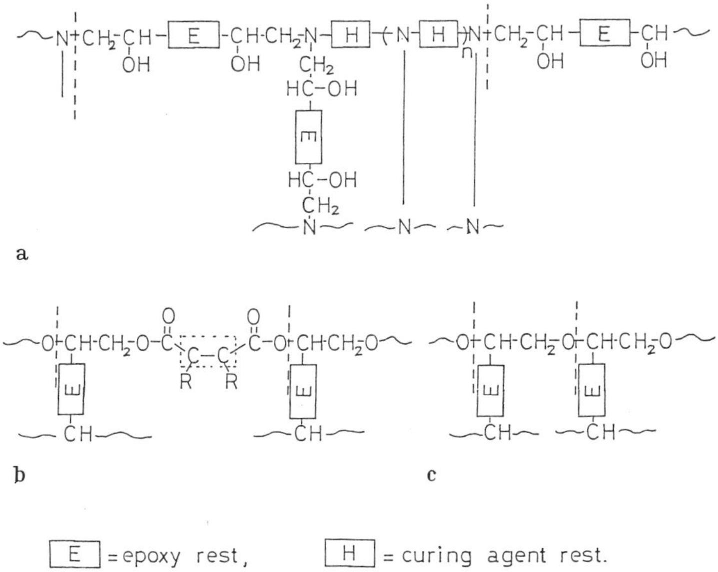

a

b              c

$\boxed{E}$ = epoxy rest,    $\boxed{H}$ = curing agent rest.

**Fig. 1a–c.** Schematic structure of fully cured epoxy resin. **a** amine cured resin; **b** anhydride cured resin; **c** catalyst cured resin

The anhydrides are sometimes used alone, but more frequently in combination with such basic compounds as tertiary amine catalysts. The suggested curing mechanism is roughly as follows [14-16]:

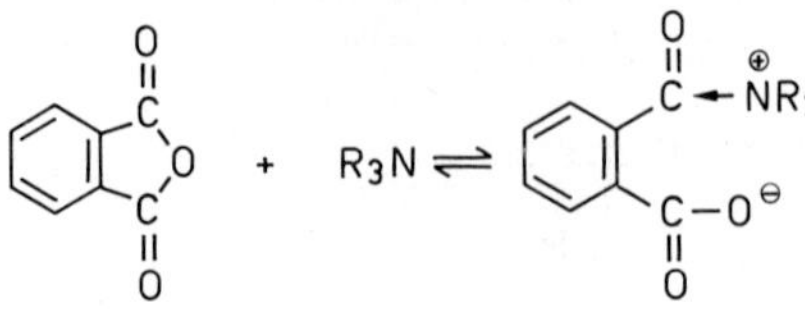

(4)

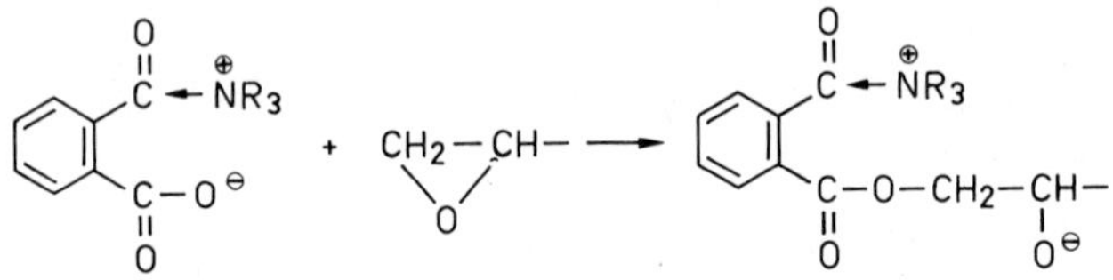

(5)

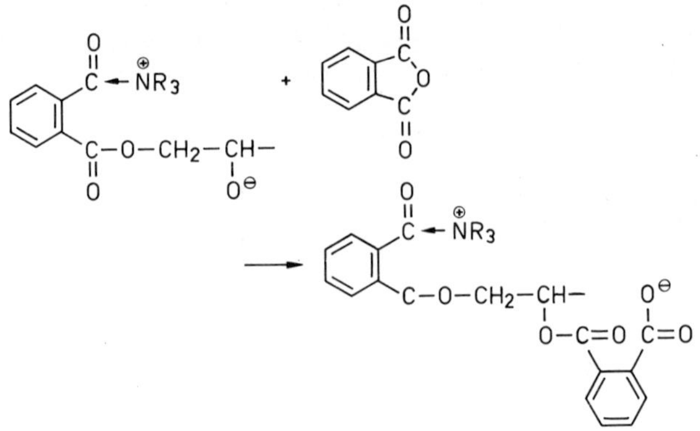

(6)

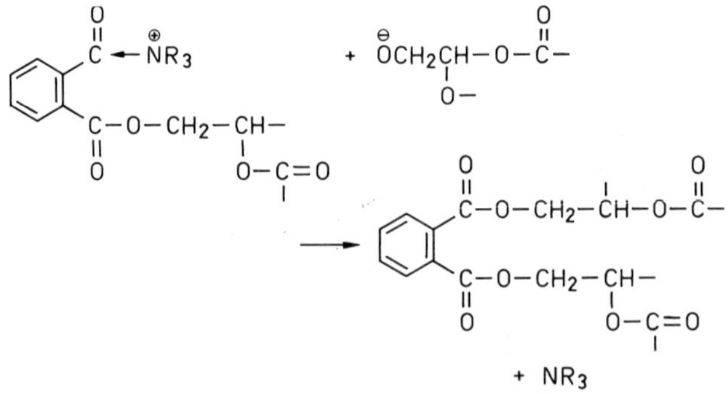

(7)

A different initiating mechanism than that given by Eq. (4) has been proposed [15-17]. The curing reaction generally proceeds as shown in Eq. (6), resulting in the alternating copolymerization of epoxy groups and acid anhydrides. The structure of the cured resin is shown in Fig. 1b. It is, however, well known that some epoxy groups may react with other epoxy groups.

Another curing mechanism is operative in the curing reaction initiated by a polymerization catalyst for epoxy ring opening, where anionic catalysts such as tertiary amines [18,19] and imidazols [20,21] and cationic catalysts such as amine complexes of

Lewis acids [22, 23] are used. Among these curing agents, many show a particular curing behavior; for example, the BF$_3$-amine complex [22, 23] is an excellent latent curing agent, and diaryliodonium salts [24] are known as agents to initiate UV curing [22–24]. Some of the initiating mechanisms are not yet clear. In general, polyethers are produced through the ring-opening polymerization of epoxy groups as shown in Eq. (8):

$$n\ R-\underset{\underset{O}{\diagdown\diagup}}{C-C} \xrightarrow{\text{Cat.}} \cdots \left(\underset{R}{C}-C-O\right)_n \underset{R}{C}-C-O-\cdots \tag{8}$$

The structure of the finally cured resin is shown in Fig. 1c.

## 3 Types of Curing Agents and the Dynamic Mechanical Properties of Cured Resins

Murayama [8] studied a series of resins from diglycidyl ether of Bisphenol A (DGEBA) cured with varying quantities of diaminodiphenylmethane (DDM). These cured resins had the same main chain structure, but differed in the degree of crosslinking.

Izumo [25] performed similar studies by using diethylenetriamine (DETA) as a

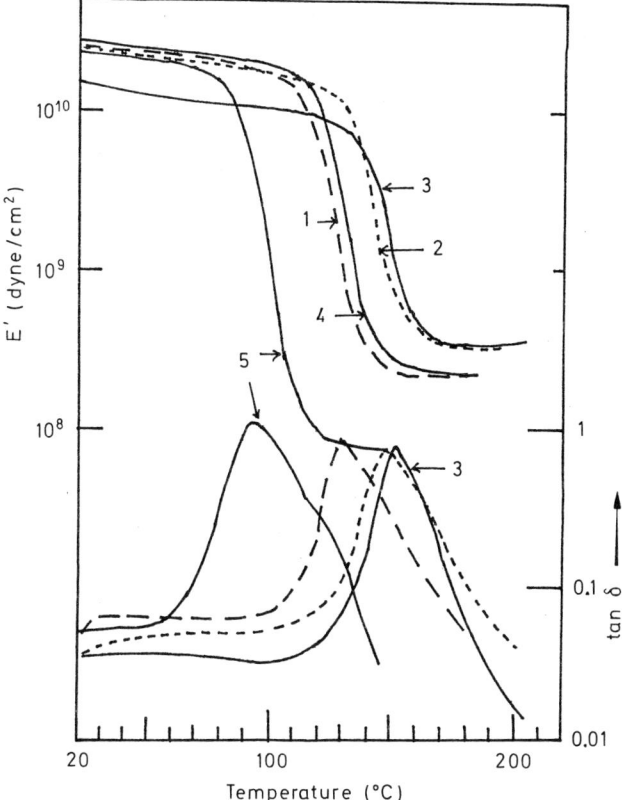

**Fig. 2.** Dynamic modulus and loss tangent vs. temperature for EDA cured epoxy resins [26]. Mol ratio (amine/epoxy); 1: 1.4, 2: 1.2, 3: 1.0, 4: 0.8, 5: 0.6

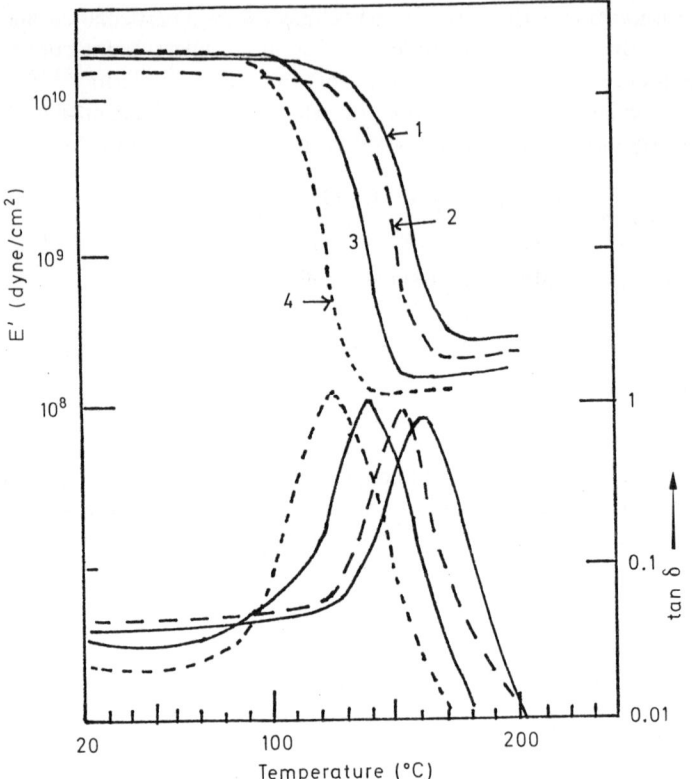

**Fig. 3.** Dynamic modulus and loss tangent vs. temperature for Me—HHPA cured epoxy resins [27]. Mol ratio (anhydride/epoxy); 1: 1.0, 2: 0.9, 3: 0.8, 4: 0.7

curing agent, and Kamon [26] used ethylenediamine (EDA). An example of the results is shown in Fig. 2.

The dynamic mechanical properties of a series of resins cured with varying quantities of an acid anhydride are shown in Fig. 3 [27].

The temperature dispersion of the dynamic mechanical properties of all these cured resins indicates three regions: (1) the glassy region where the dynamic Young's modulus ($E'$) is about $10^{10}$ dyne/cm$^2$ which changes little at lower temperature, (2) the transition region where $E'$ is changed from $10^{10}$ to $10^8$ dyne/cm$^2$ in a narrow temperature range and (3) the rubbery region where $E'$ becomes equal to about $10^8$ dyne/cm$^2$ and is constant at higher temperatures. The temperature indicating the maximum tan $\delta$ is taken as the glass transition temperature $T_g$.

The value of $E'$ in the rubbery region was theoretically and experimentally studied and it was found to be proportional to the crosslinking density ($\varrho$) or inverse to the molecular weight between linkings ($M_c$). Tobolsky [1, 28], used Eq. (9) to demonstrate the interrelationship:

$$E' = 3\varphi \, dRT/M_c \tag{9}$$

If the effect of dangling is neglected, then:

$$M_c = d/\varrho$$

Thus, Eq. (9) changes to:

$$E' = 3\varphi\varrho RT \qquad (10)$$

Here, E' is the Young modulus in the rubbery state, d the density, R the gas constant, T the temperature in K, $M_c$ the molecular weight between crosslinks, $\varrho$ the crosslinking density, and $\varphi$ the front factor. The value of $\varphi$ is close to unity.

## 3.1 Dynamic Mechanical Properties of Cured Resins with Different Curing Mechanisms

Although the cured epoxy resins produced by the three typical curing mechanisms have different structure, the temperature dependence of the dynamic mechanical dispersion is similar as shown in Figs. 2 and 3, but $T_g$ is different. Therefore, in order to check the relationship between the structure and $T_g$, $\varrho$ should be identical.

The relationship between $T_g$ and $\varrho_{(E')}$ ($\varrho_{(E')}$ has been calculated from Eq. (9) for $\varphi = 1$) of a series of the cured resins with different $\varrho$ are shown in Fig. 4. The curing agents used are DETA [25], EDA [26], and DDM [8] as a diamine, succinic acid anhydride (Suc A) [27] as an acid anhydride and 2-ethyl-4-methyl imidazole (Im) [21] as a catalytic curing agent.

The relationships between $T_g$ and $\varrho$ have been studied in a number of papers [29-31]. Shibayama studied theoretically the relationships based on the volume shrinkage due to crosslinking, and suggested that Eq. (11) was applicable to many crosslinked high polymers [32].

$$T_g = K_1 \log K_2 \varrho \qquad (11)$$

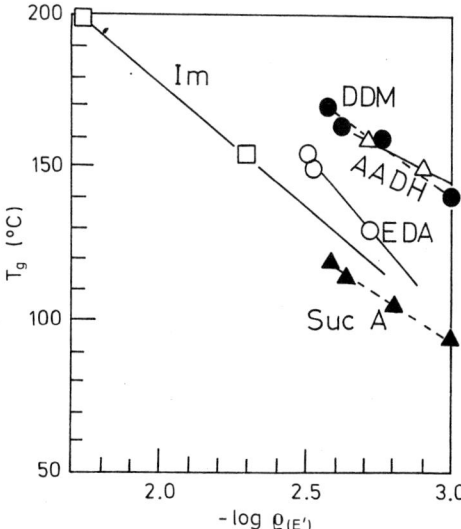

Fig. 4. Glass transition temperature ($T_g$) vs. crosslinking density ($\varrho_{(E')}$) for resins cured with various curing agents

**Table 1.** $K_1$ and log $K_2$ in Eq. (11) for epoxy resins cured with various curing agents

| Curing agent | $K_1$ | log $K_2$ |
|---|---|---|
| EDA | 112.7 | 3.86 |
| DETA | 96.3 | 3.84 |
| DDM | 59.3 | 5.5 |
| AADH | 51.0 | 5.92 |
| SucA | 50.4 | 3.68 |
| Me-HHPA | 84.8 | 4.56 |
| PnEThGE | 47.3 | 4.61 |
| TETA-MeDPTA | 107 | 3.82 |
| EDA-HMDA | 216 | 3.23 |
| AADH-SL-20DH | 248 | 3.36 |
| Polyanhydride | 227 | 3.21 |
| Imidazole | 78.6 | 4.27 |

Resin: DGEBA

Here, $K_1$ and $K_2$ are constants for the same groups of polymers; $K_2$ depends on the strength of interaction and the rigidity of the main chain, and $K_1$ is related to the volume shrinkage due to crosslinking and characterizes the degree of restraint of the molecular motion near the crosslink ($K_1$ is about 100 for no restraint, and approaches zero as the restraint increases).

The dependence predicted by Eq. (11) is obeyed for a number of epoxy resin-curing agent systems, as shown in Fig. 4. The values of $K_1$ and log $K_2$ shown in Table 1.

The values of log $K_2$ are nearly the same, which indicates that the resins are almost the same in physical properties, although their main chain structures are different (Fig. 1).

$K_1$ is nearly 100 for EDA and DETA since there is no restraint in the molecules of the curing agents; $\varrho_{(E')}$ is higher and $T_g$ is slightly elevated in Im and Suc A since the restraint is high. This is due to increasing flexibility of the main chain in the series: hydrocarbon < polyether < polyester. For DGEBA resins, $T_g$ increases in the series: amine- > catalyst- > acid anhydride, for the same value of $\varrho_{(E')}$.

If in the H segment in Fig. 1 the alkyl group is replaced by an aryl group, e.g., EDA by DDM, or Suc A by methylhexahydrophthalic anhydride (MeHHPA), $K_2$ becomes higher, the rigidity of the main chain increases, and $T_g$ becomes higher (Table 1).

## 3.2 Structures and Dynamic Mechanical Properties of Epoxy Resins Cured with Polyamines

Amines are the most frequent curing agents, and the curing mechanism is determined by a simple addition reaction of active hydrogens of polyamines to epoxy groups. Consequently, the structures of the cured resins are not much different from the ideal structure shown in Fig. 1a. Some examples of polyamines are shown in Table 2. The structures of the cured resins obtained from DGEBA and stoichiometrically equivalent amounts of polyamines are shown in Fig. 1a, and differ in the H segment.

Table 2. Structure and boiling point of polyamine curing agents

| Code | Structure | b.p. (°C) |
|---|---|---|
| EDA | $H_2N(CH_2)_2NH_2$ | 117 |
| DETA | $H_2N(CH_2)_2NH(CH_2)_2NH_2$ | 208 |
| TETA | $H_2N(CH_2)_2NH(CH_2)_2NH(CH_2)_2NH_2$ | 157 °C/20 mm |
| PDA | $H_2N(CH_2)_3NH_2$ | 140 |
| DPTA | $H_2N(CH_2)_3NH(CH_2)_3NH_2$ | 241 |
| MeDPTA | $H_2N(CH_2)_3N(CH_3)(CH_2)_3NH_2$ | 234 |
| HMDA | $H_2N(CH_2)_6NH_2$ | 205 (m.p. 42 °C) |
| TMAH | $H_2N(CH_2)_3CH(CH_2NH_2)(CH_2)_3NH_2$ | 160 °C /10 mm |
| MXDA | $H_2NCH_2\text{-}C_6H_4\text{-}CH_2NH_2$ | 245 °C |
| LARO | $H_2N\text{-}C_6H_9(CH_3)\text{-}CH_2\text{-}C_6H_9(CH_3)\text{-}NH_2$ | 200~212 °C / 20 mm Hg |
| DDM | $H_2N\text{-}C_6H_4\text{-}CH_2\text{-}C_6H_4\text{-}NH_2$ | (m.p. 90 °C) |

The temperature dependence of the dynamic mechanical properties of the cured resins is similar as in Fig. 5 [33]. This shows that the physical properties of cured resins are mostly characterized by the differences of $T_g$.

The diamine curing agents $H_2N(CH_2)_nNH_2$ ($n = 2, 3, 4, 6$) have been taken as an example (cf. Table 2).

It can be assumed that $T_g$ is lowered since $M_c$ becomes higher with increasing n [34] [Fig. 5, Eqs. (10) and (11)].

As n increases from 2 to 6, $\varrho_{(E')}$ is lowered and the rigidity of the main chain decreases. It is suggested that the effect of this decrease on $T_g$ is almost constant [34].

The following example shows that $M_c$ can be simply changed by changing the functionality of the hardeners which is 6, 5, and 4 for TETA, DPTA and MeDPTA, respectively. The relationships between $T_g$ and log $\varrho_{(E')}$ are described by straight lines in Fig. 6, where $K_1$ is 107 which is similar to the value found for EDA and DETA systems, and show clearly the effect of the phenyl groups on $T_g$ (Table 1).

The resins cured with EDA, DETA and TETA were already studied by Katz [35]. In this system, the number of the functional groups of the polyamine is increased if

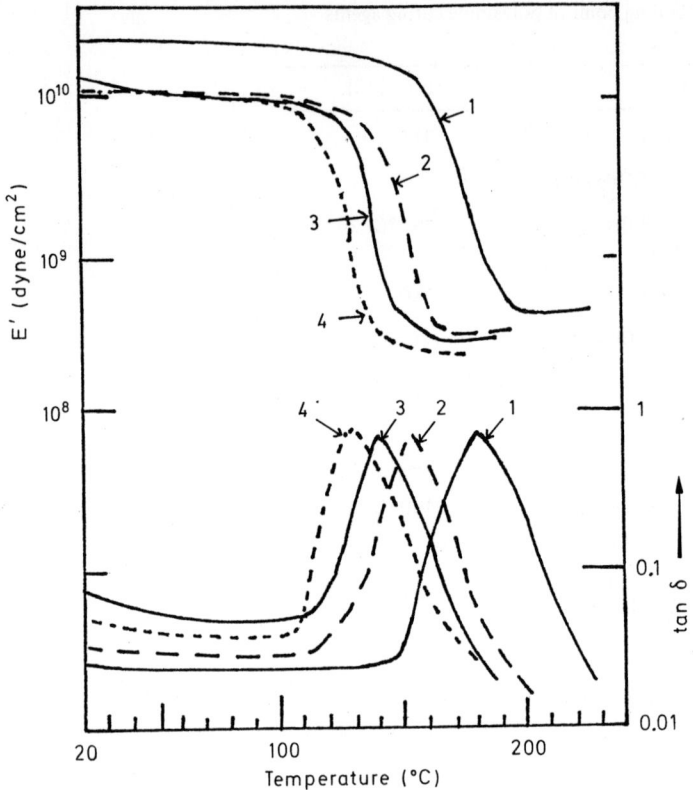

**Fig. 5.** Dynamic modulus and loss tangent vs. temperature for various diamine cured epoxy resins [33]. 1: DDM, 2: EDA, 3: BDA, 4: HMDA

EDA is replaced by TETA. Accordingly, despite the increase of $\varrho_{(E')}$, $T_g$ is slightly lowered [34] (cf. Fig. 6).

TMAH is a diamine having a primary amine group in the center of the molecule, and the dynamic mechanical properties of the cured resin are interesting (cf. Table 3). Both $T_g$ and $\varrho_{(E')}$ are almost the same as for TETA, which suggests that such a slight

**Table 3.** Dynamic mechanical properties and flexural strength of diamine cured epoxy resins

| Curing agent | $T_g$ (°C) | $E' \times 10^{-10}$ (dyne/cm²) | $\varrho_{(E')}$ $\times 10^3$ (mol/cm³) | Flexural strength (Kg/cm²) |
|---|---|---|---|---|
| DDM | 180 | 2.52 | 2.51 | 1230 |
| LARO | 180 | 1.70 | 2.34 | 1070 |
| MPDA | 185 | 1.86 | 2.40 | 1090 |
| MXDA | 135 | 2.01 | 1.71 | 1240 |
| BDA | 140 | 1.29 | 2.28 | 750 |

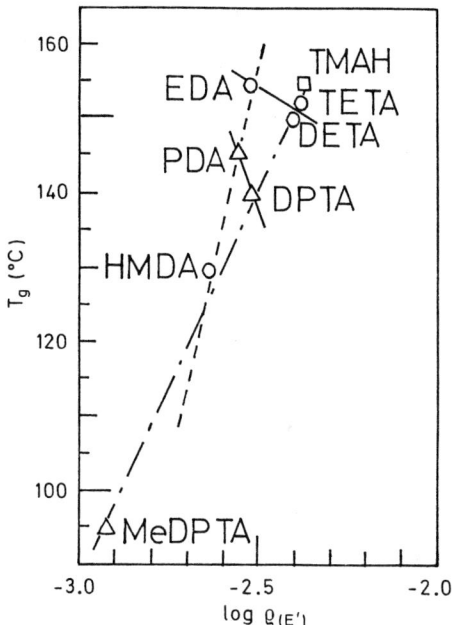

**Fig. 6.** $T_g$ vs. log $\varrho_{(E')}$ for aliphatic polyamine cured epoxy resins [34]

difference in molecular structure has almost no influence on physical properties [34].

If an alkyl chain (BDA) is substituted by *m*-phenylenediamine (MPDA), $T_g$ is elevated by 45 °C (Table 3). However, almost the same effect is observed for DDM, while for MXDA having an alkyl-phenyl group, almost no effect on $T_g$ is observed because the rigidity of the main chain is not increased enough.

In both cases of cycloaliphatic and phenyl groups, nearly the same effects on the increase of $T_g$ are observed. LARO and DDM can serve as example (cf. Table 3) [34].

## 3.3 Dynamic Mechanical Properties of Resins Cured with Carboxylic Hydrazides

In Section 3.2, the effect of the segment structure and the number of functional groups on physical properties is discussed. There, it has been analyzed how the structure of the crosslinks $(-N\langle)$ influences the physical properties. Since carboxylic hydrazides can be easily synthesized from carboxylic acids, hardeners of various structures can be obtained. As the structure of the functional group $(-CONHNH_2)$ is similar to that of amines, it is interesting to compare the properties of the cured resins. The hydrazides are powders with high melting points and are used as latent curing agents with a pot life of more than four months [36, 37] (Table 4).

The curing mechanism of epoxy resins with carboxylic hydrazides has not been sufficiently revealed so far. However, Kamon [36] has shown that one hydrazide group reacts with two epoxy groups. This result suggests that the structure of the epoxy resin cured with hydrazide is presumed to be nearly the same as that shown in Fig. 1a, where the $-N\langle$ segment is substituted by the $-CONHN\langle$ segment.

**Table 4.** Structure and melting point of dihydrazides

| Code | H$_2$NHNCO—R—CONHNH$_2$<br>R | m. p.<br>(°C) |
|---|---|---|
| SuADH | $\text{-(CH}_2\text{)}_2\text{-}$ | 163 |
| AADH | $\text{-(CH}_2\text{)}_4\text{-}$ | 180 |
| SeADH | $\text{-(CH}_2\text{)}_8\text{-}$ | 184~5 |
| DDADH | $\text{-(CH}_2\text{)}_{10}\text{-}$ | 184~5 |
| SL-16DH | $\text{-(CH}_2\text{)}_{14}\text{-}$ | 174 |
| IPADH | ⌬ | 212 |
| N-PDPDH | —CH$_2$CH$_2$—N—CH$_2$CH$_2$—<br>(Ph) | 141~5 |

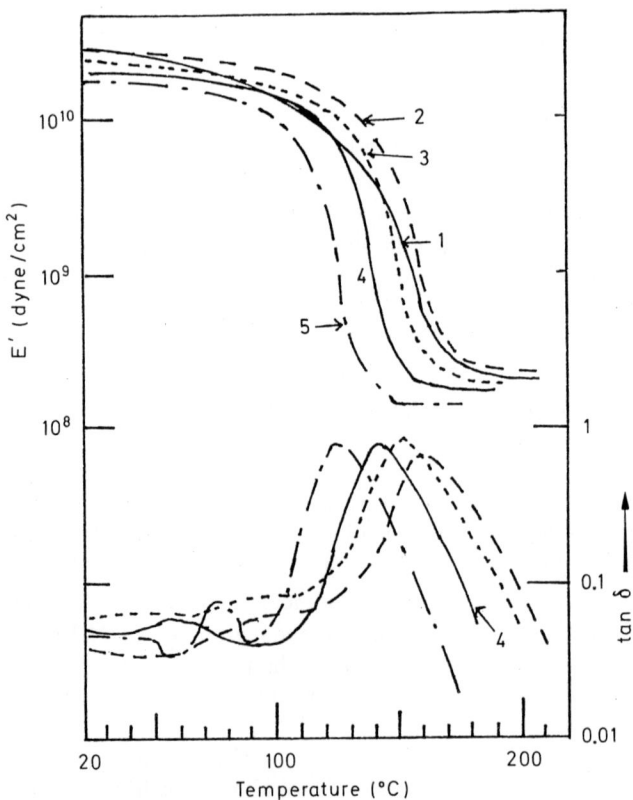

**Fig. 7.** Dynamic moduls vs. temperature for aliphatic dihydrazide cured epoxy resins [37]. 1: SuADH, 2: AADH, 3: SeADH, 4: SL-16DH, 5: SL-20DH

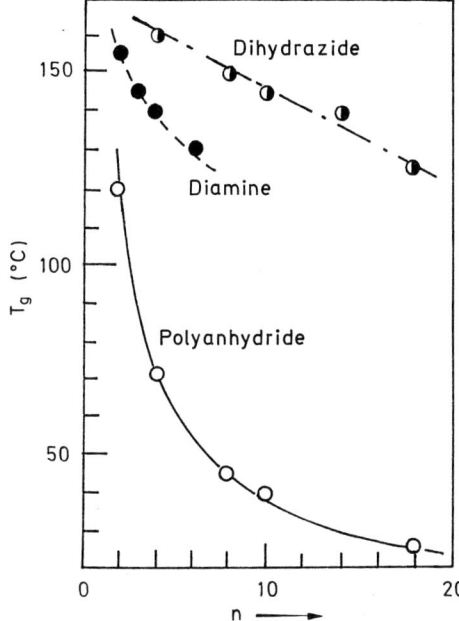

Fig. 8. Effect of the number of methylene group (n) in diamine, dihydrazide and polyanhydride for cured epoxy resins on $T_g$.[41]

The temperature dependence of the dynamic modulus of the hydrazide-cured resins is very similar to that of amine-cured resins as shown in Fig. 7. Based on the dynamic mechanical properties of a series of the resins with different $\varrho_{(E')}$ and cured by changing the amount of AADH, $T_g$ and log $\varrho_{(E')}$ are plotted in Fig. 4. In this case, $T_g$ is by 30 °C higher than $T_g$ of the resin with the same $\varrho_{(E')}$ cured by ethylenediamine (EDA). This difference is due to the polarity of the —CONHN— segment in the main chain which is higher than that of the —N— segment. This is corroborated by the fact that $K_2$ obtained by Eq. (11) is larger than that for the alkyldiamine-cured resin, and is nearly the same as that of the DDM-cured resin (Table 1).

The results in the series of resins cured with a series of $\alpha,\omega$-alkylene dihydrazides again show that $T_g$ decreases with increasing number of $CH_2$ groups (Fig. 8). However, $T_g$ of hydrazide-cured resin is higher than that of the corresponding alkylenediamine-cured resin due to polarity differences.

The relation between $T_g$ and $\varrho_{(E')}$ in Eq. (11) is linear as shown in Fig. 9, and $K_1$ and log $K_2$ values are very close to those of EDA-HMDA-cured resins.

As presumed from those facts, $T_g$ of a resin cured with IPADH, in which the phenyl groups have replaced the $CH_2$ groups, is higher than that of the AADH-cured resin and reaches 180 °C [37].

In addition, $T_g$ of the N-PDPDH-cured resin having phenyl side group is 165 °C, i.e., by 5 °C higher than that for the AADH-cured resin having nearly the same number of atoms between functional groups. This indicates that the phenyl groups, even as side groups, are effective in rising $T_g$ [37]. The presence of phenyl groups in the resin structure not only makes the main chain more rigid but also decreases the free volume.

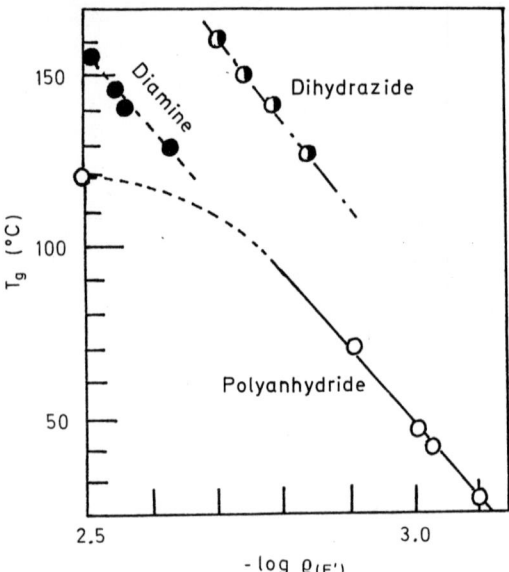

Fig. 9. $T_g$ vs. $\varrho_{(E')}$ for polyanhydride, Diamine and dihydrazide cured epoxy resins [41]

## 3.4 Dynamic Mechanical Properties of Epoxy Resins Cured with Carboxylic Anhydride

Curing with anhydride hardeners requires a long curing time and a high temperature. The cured resins are excellent in mechanical and particularly in electrical properties at temperatures above $T_g$. Anhydride hardeners are extensively used. Many practical studies on mechanical and electrical properties [38] but few systematic studies on the structures of the cured resins and physical properties have been reported [27, 39–41]. The reasons why such systematic studies are scarce are as follows: Most of the acid anhydride hardeners are cyclic anhydrides of dibasic acids. The main structure of the cured resins is such as shown in Fig. 1b. Differences among the structure of the cured resins with different hardeners are only based on the differences in the R segment and consequently, cured resins with rather different mechanical properties cannot

**Table 5.** Structure and melting point of polyanhydrides

| Code | HO$-$$\{$O$-$C$-$(CH$_2$)$_n$$-$COO$\}_m$$-$H | | m. p. (°C) |
|---|---|---|---|
| | $n$ | $m$ | |
| SucAA | 2 | ($\infty$) | 120 |
| PAPA | 4 | 2.1 | 77.5–86 |
| PSPA | 8 | 2.5 | 73–83 |
| SL-12 AH | 10 | 3.19 | 75–80 |
| SL-20 AH-2 | 18 | 3.16 | 82–87 |
| (MeHHPA[a]) | 2 | ($\infty$) | liquid |

[a] MeHHPA = methyl-hexahydrophthalic anhydride

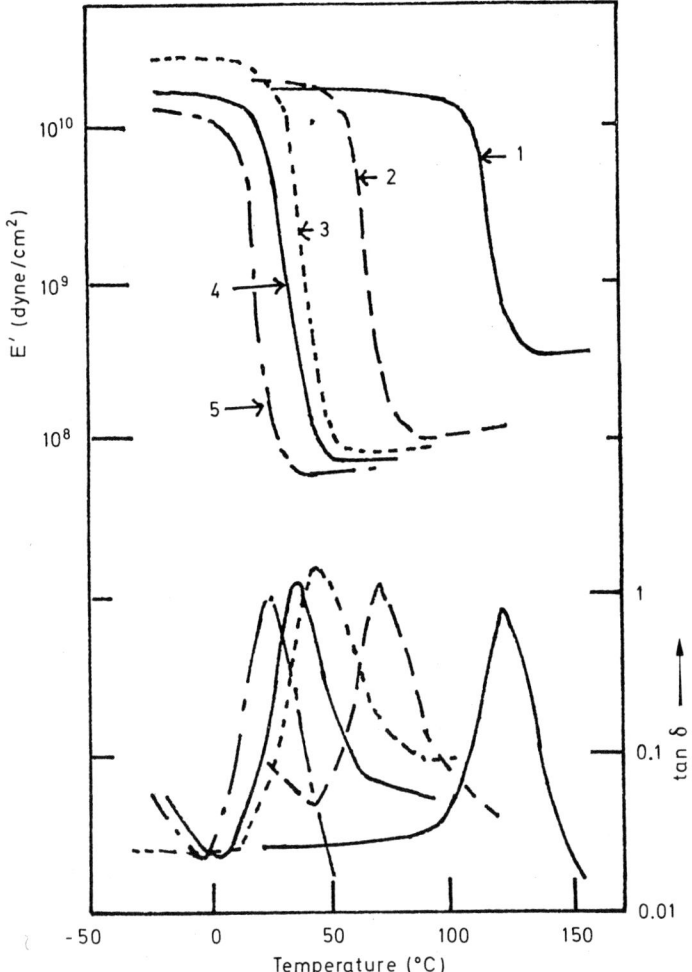

**Fig. 10.** Dynamic modulus (E') and loss tangent vs. temperature for various polyanhydride cured epoxy resins (10 Hz) [41]. 1: Suc A, 2: PAPA, 3: PSPA, 4: SL-12AH, 5: SL-20AH

be expected. Also, the curing mechanism is complicated as described in chapter 2, and there is a pronounced difference between the ideal structure shown in Fig. 1b and the actual structure of cured resin.

Recently, the dynamic mechanical properties of the resins cured with a series of linear polyanhydrides having the structures shown in Table 5 were studied by Kamon [41]. The dynamic mechanical properties are very similar to those of the resins cured with diamine hardeners as shown in Fig. 10. The relationships between the number of $CH_2$ groups n and $T_g$ are shown in Fig. 8. $T_g$ drops rapidly below that of an alkyldiamine-cured resin as n increases. This is considered reasonable because, compared with the hardener unit in an amine-cured resin, the acid anhydride segment existing in the main chain of the anhydride-cured resin is twice as long.

**Table 6.** Dynamic mechanical properties for epoxy resin cured with acid anhydrides

|  | PA | MNA | Me—HHPA | HHPA | Suc A | MA-1 | MA-2 |
|---|---|---|---|---|---|---|---|
| $T_g$ (°C) | 160 | 160 | 160 | 155 | 120 | 130 | 150 |
| $d$ ($T_g$ + 40 °C) | 1.231 | 1.170 | 1.171 | 1.168 | 1.222 | 1.233 | 1.267 |
| $E'$ (dyn/cm$^2$) (20 °C) × 10$^{10}$ | 2.408 | 2.614 | 1.923 | 1.811 | 1.711 | 1.776 | 2.089 |
| $E'$ (dyn/cm$^2$) ($T_g$ + 40 °C) × 10$^{-8}$ | 2.996 | 2.321 | 2.601 | 2.236 | 3.453 | 2.203 | 4.076 |
| $\varrho_{(E')} = E/3dRT$ (mol/g) × 10$^3$ | 2.062 | 1.681 | 1.882 | 1.639 | 2.615 | 1.616 | 2.785 |

Those relations between $T_g$ and log $\varrho_{(E')}$ are nearly linear except of Suc A, as shown in Fig. 9. The values of $K_1$ and $K_2$ derived from Eq. (11) are very close to those for the diamine and dihydrazide-cured resins, as shown in Table 1. As $\varrho_{(E')}$ becomes smaller, the differences in main chain structures have almost no effect on $T_g$, but only the increase of $\varrho_{(E')}$ can increase $T_g$.

However, at higher crosslinking density, $T_g$ increases in the series dihydrazide > diamine > anhydride; $T_g$ is influenced by the polarity and flexibility of the main chain.

The dynamic mechanical properties of aromatic and cycloaliphatic acid anhydrides are very similar, because they differ only in the R segment as shown in Fig. 1b. The $T_g$ of cured resins containing phenyl or cycloaliphatic groups is higher than that of Suc A, but lower than $T_g$ of amine- or hydrazide-cured resins (Table 6). This is because the flexibility of the main chain of the acid anhydride-cured resin is high, and there are no OH groups available and hydrogen bonding is not operative.

It is interesting that the difference between Suc A ($T_g$ = 120 °C) and BDA ($T_g$ = 140 °C) is nearly the same as between phthalic anhydride (PA) ($T_g$ = 160 °C) and MPDA ($T_g$ = 185 °C) or DDM ($T_g$ = 180 °C).

## 3.5 Dynamic Mechanical Properties of Mercaptan-cured Resin

The mercaptan hardener is one of a very few hardeners which can cure the resin at low temperature even below the freezing point in a few minutes using basic catalysts such as amines. There are only very few studies of the curing reaction mechanism and the physical properties of the cured resin available [42-45].

The mechanism of curing of epoxy resin with mercaptan has been proposed by Kamon [44] and has been based on the work by Danehy [46]:

$$R-SH + NR'_3 \rightleftarrows RS^\ominus + H^\oplus NR'_3 \tag{12}$$

$$RS^\ominus + \underset{O}{C-C-R''} \rightarrow RS-C-C-R'' \quad\quad\quad (13)$$
$$\phantom{RS^\ominus + C-C-R'' \rightarrow RS-C-}\underset{O^\ominus}{|}$$

$$RS-C-\underset{O^\ominus}{\overset{|}{C}}-R'' + R-SH \rightarrow RS-C-\underset{OH}{\overset{|}{C}}-R'' + RS^\ominus \tag{14}$$

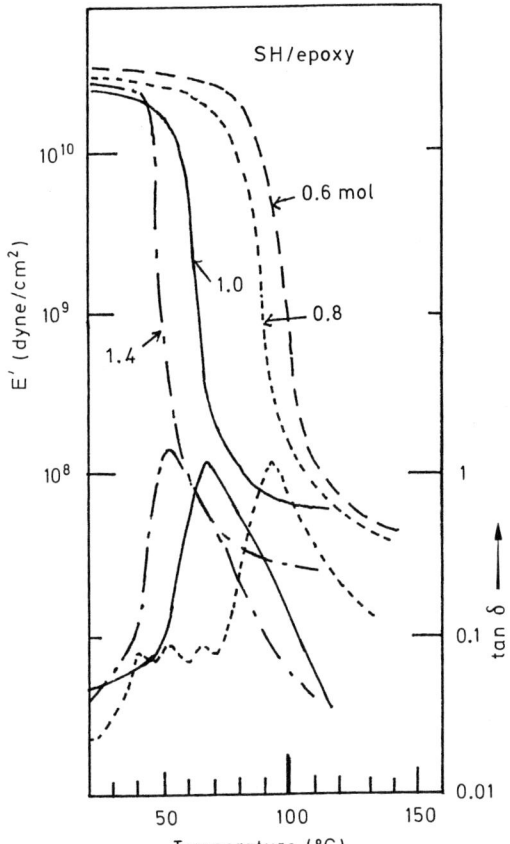

**Fig. 11.** Dynamic modulus and loss tangent vs. temperature for PnEThGE cured epoxy resins [44]

Accordingly, the structure of the cured resin with polymercaptans having more than three functional groups would be close to that shown in Fig. 1a. Thioglycolate of pentaerythritol (PnEThGE) (structure [I] below) is a hardener with four functional groups, and the physical properties of the cured resin can be compared with those cured with diamine or dihydrazide.

$$\text{HSCH}_2\text{COOCH}_2-\underset{\underset{\text{CH}_2\text{OCOCH}_2\text{SH}}{|}}{\overset{\overset{\text{CH}_2\text{OCOCH}_2\text{SH}}{|}}{\text{C}}}-\text{CH}_2\text{OCOCH}_2\text{SH} \tag{I}$$

The dynamic mechanical properties of the cured resin are very similar to those of resins cured with amines or acid anhydrides (Fig. 11) [44].

However, by changing the amount of hardener $\varrho_{(E')}$ passes through a maximum as shown in Fig. 12. The maximum corresponds to the equivalent ratio of SH and epoxy groups. $T_g$ is higher when epoxy groups are in excess, which differs from the case when other hardeners are used in curing. An excess of epoxy groups polymerizes

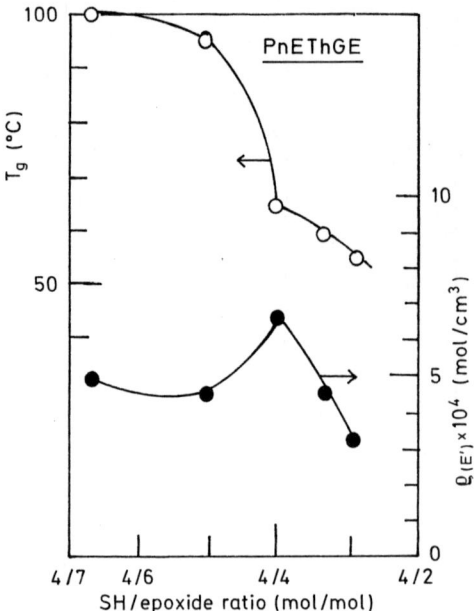

Fig. 12. Effect of polymercaptan concentration of cured DGEBA on $T_g$ and $\varrho_{(E')}$ [44]

easily, so that the structure of the cured resin contains both —S— and —O— bridges.

The relationship between $T_g$ and log $\varrho_{(E')}$ in the cured resin with an excess of SH groups is linear. $K_1$ and log $K_2$ obtained by Eq. (11) are 47 and 4.6, respectively (Table 1).

The value of $K_1$ is smaller than that of alkyldiamine-cured resin, because the resin contains ester groups (structure [I]) of the hardener, but mainly because its cross-linking segments are longer than those of amines [44].

## 4 Dynamic Mechanical Properties of Resins Cured with Different Hardeners

The structure of the cured epoxy resin cannot be always simply derived from the structure of starting components but one has to consider the curing mechanism. A few examples are given below.

### 4.1 Dicyandiamide

Dicyandiamide (DICY) is a powder with a high melting point [II], and is suitable for excellent latent one-component liquid resin systems.

$$H_2N-\underset{\underset{NH}{\|}}{\overset{\overset{H}{|}}{C}}-N-C\equiv N \tag{II}$$

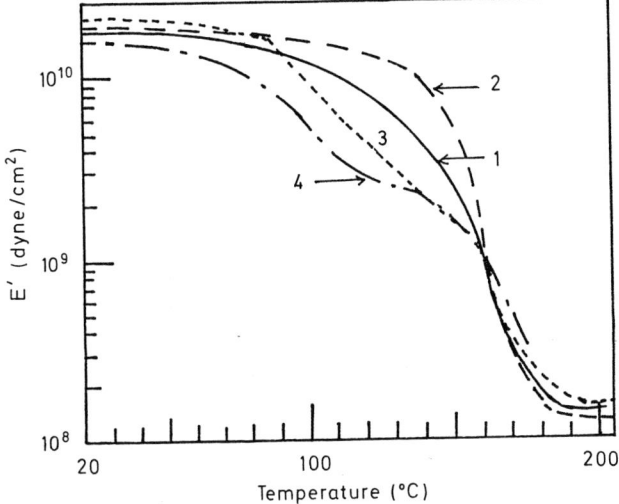

**Fig. 13.** Effect of DICY concentration on E'-temperature curve (curing on 2 hr at 170 °C) (10 Hz) [47]. 1: 1.5 eq., 2: 1.0 eq., 3: 0.45 eq., 4: 0.25 eq.

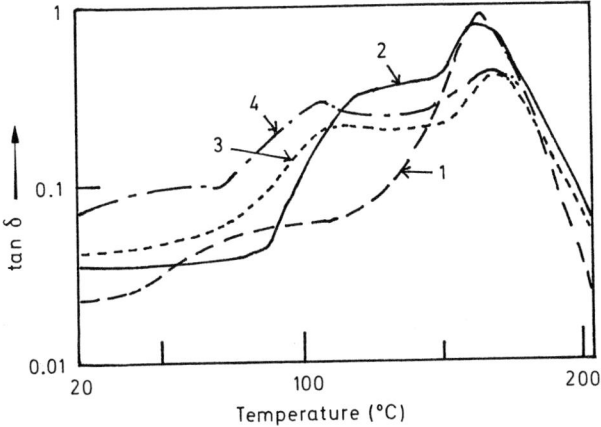

**Fig. 14.** Effect of DICY concentration on loss tangent-temperature curve [47]. 1: 1.0 eq., 2: 0.8 eq., 3: 0.45 eq., 4: 0.25 eq.

The structure of DICY resembles that of diamine hardeners but the dynamic mechanical properties of the resin cured with DICY is far different from that of the diamine-cured resin in Fig. 2 (Figs. 13 and 14) [47].

The resin cured with excess DICY (assuming DICY is tetrafunctional) is a single phase system while the resins cured with DICY with epoxy groups in excess are two-phase systems which exhibit two maxima in tan δ. Since the IR analysis shows that all NH and epoxy groups disappear, the main reaction is assumed to be the addition reaction of the epoxy groups to the NH groups.

However, $\varrho_{(E')}$ obtained from E' in the rubbery region is $1.2 \times 10^{-3}$ mol/cm³, which is only about 1/3 of $\varrho_{(E')}$ of the EDA cured resins ($3.3 \times 10^{-3}$ mol/cm³) (cf. Fig. 2). From these facts and taking into account the results of Saunders [48], Kamon et al. [47] proposed the following curing mechanisms

$$H_2N-\underset{\underset{NH}{\|}}{C}-NH-C\equiv N \rightleftarrows 2\,H_2N-C\equiv N \qquad (15)$$

$$H_2N-CN + 2\,R-CH-CH_2 \rightarrow R-\underset{OH}{\overset{}{C}H}-CH_2-\underset{CN}{\overset{}{N}}-CH_2-\underset{OH}{\overset{}{C}H}-R \qquad (16)$$
$$\phantom{H_2N-CN + 2\,R-CH}\underset{O}{\diagdown\diagup}$$

$$\begin{array}{c}
R'-\underset{CN}{\overset{}{N}}-R' \\
+ \\
\underset{\underset{CN}{\overset{}{|}}}{\overset{OH}{\overset{}{|}}}{R-CH-CH_2-N-R'}
\end{array}
\longrightarrow
\begin{array}{c}
R'-\underset{\underset{\underset{R-CH-CH_2-N-R'}{\overset{O}{\overset{}{|}}}}{\overset{C=NH}{\overset{}{|}}}}{\overset{}{N}}-R' \\
\phantom{R'-N} \downarrow\uparrow \phantom{CN}\\
R'-\underset{\underset{\underset{R-CH-CH_2-N-R'}{\overset{NH}{\overset{}{|}}}}{\overset{C=O}{\overset{}{|}}}}{\overset{}{N}}-R' \\
\phantom{R-CH-CH_2-N-R'}\underset{CN}{\phantom{|}}
\end{array}
\qquad (17)$$

R: Epoxy group
R': R—CHOH—CH₂—

In the cured resin described by Eq. (17), $\varrho_{(E')}$ is smaller, but the crosslinking segment is shorter and its polarity higher due to —CONH—; consequently, $T_g$ becomes higher [47]. The two-phase system obtained in case of excess epoxy groups is complicated. Apparently, the epoxy groups remaining after completion of the addition polymerize to yield the structure in Fig. 1c which constitutes the second phase [47].

## 4.2 Resins Cured with Isocyanates

By using isocyanates as hardeners, more complicated cured resins are obtained. It is well known that poly-2-oxazolidone is obtained by the reaction of a diepoxide and a diisocyanate compound [49, 50].

It has recently been reported that in the system epoxy resin/diisocyanate with imidazole as catalyst, the diisocyanate trimerizes at a temperature below 100 °C to form an isocyanurate ring, while above 110 °C the epoxy group and isocyanate group react to form oxazolidone, and a three-dimensional network [Eq. (18)] is obtained [51].

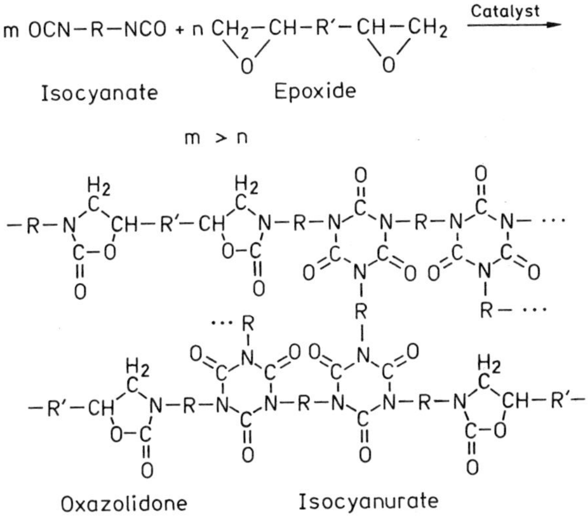

$$m\ OCN-R-NCO + n\ CH_2-CH-R'-CH-CH_2 \xrightarrow{Catalyst}$$

Isocyanate    Epoxide

$m > n$

[Structural diagram showing Oxazolidone and Isocyanurate units]    (18)

By changing the ratio epoxy/isocyanate, the cured resins are composed of oxazolidone and isocyanurate rings in a varying ratio.

The dynamic mechanical properties of three epoxy resins cured with diphenylmethane diisocyanate (MDI) are shown in Fig. 15. Since these resins consist of many bulky cyclic structures, $T_g$ (α-dispersion) is above 200 °C and the transition region is wide. In the DEN 431—L-MDI resin, $T_g$ is higher than 300 °C. The resin is expected to be highly heat-resistant [51].

## 4.3 Other Examples

4,4'-methylene bis(phenyl cyanamide) (MBPC) also yields cyclic structures [52]. MBPC forms 2-imino-oxazolidine rings by the reaction with epoxy groups as shown in Eq. (19). The formed imino groups (=NH) further react with epoxy groups. It is known that this is an excellent heat-resistant resin for FRP, but its dynamic mechanical properties have not yet been determined [52].

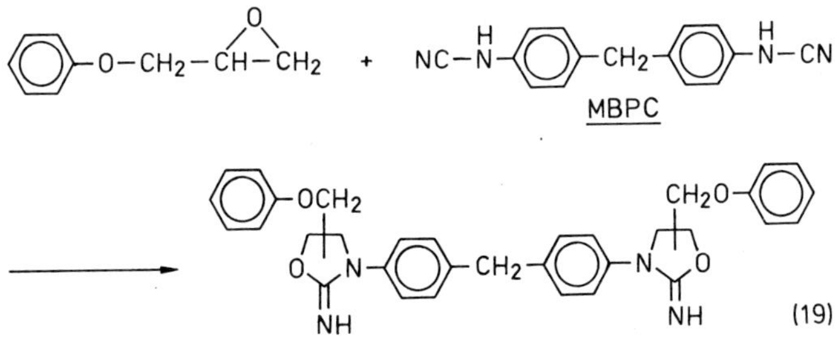

[Reaction scheme showing epoxy + MBPC giving product with 2-imino-oxazolidine rings]    (19)

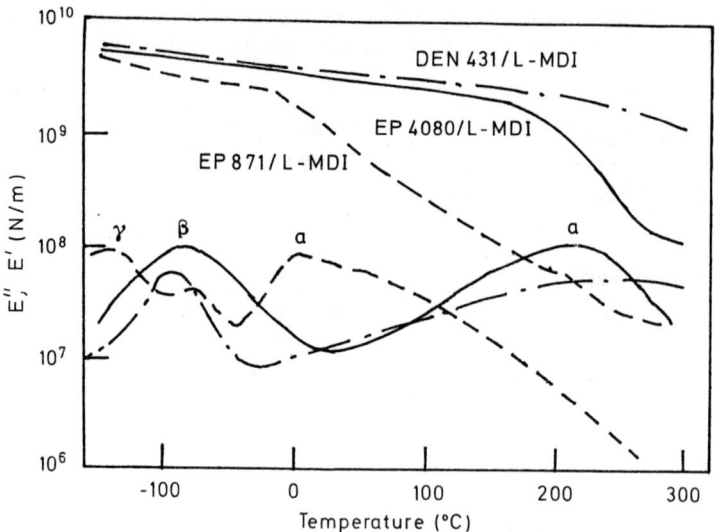

**Fig. 15.** Viscoelastic behavior of isocyanurate-oxazolidone resins [51]. DEN 431 (polyglycidyl ether of phenol-formaldehyde novolac; Dow Chemical Co.), EP 4080 (2,2-bis[p-(2,3-epoxypropyloxy)cyclo-hexane]propane; Asahi Denka, Ltd.), EP 871 (diglycidyl ester of linoleic dimer acid; Shell Chemical Co.), and L-MDI (modified diphenylmethane-4,4'-diisocyanate; Desmodur CD: Bater AG)

## 5 Mechanical Properties of Cured Epoxy Resins in the Glassy State

In the preceding sections, $T_g$ and the structures which determine the properties of the cured resins were discussed. These cured resins can be exploited as structural materials at temperatures below $T_g$.

It is known that in the glassy state below the glass transition temperature the physical properties of epoxy resin as well as other amorphous polymers are generally little dependent on temperature and structure [1, 28]. Also, the modulus of elasticity (E) does only weakly depend on the crosslinking density.

It has been reported by Bell that for the resins cured with different amounts of DDM, E at 20 °C remains the same even when $\varrho$ becomes smaller (Fig. 16) [53]. Similar conclusions have been obtained for higher-molecular-weight DGEBA resins [12, 61].

The modulus of elasticity in the glassy state depends on intermolecular forces [1]. Therefore, Tobolsky [28] proposed the following equation correlating the bulk modulus of elasticity (B) at 0 K and cohesive energy density ($\delta^2$):

$$B = 8.04\delta^2 \tag{20}$$

where $\delta$ is the solubility parameter. The relation of the modulus of elasticity (E) and $\delta$ has been derived from the data by Holliday [54]:

$$E = 13.3\delta^2 \tag{21}$$

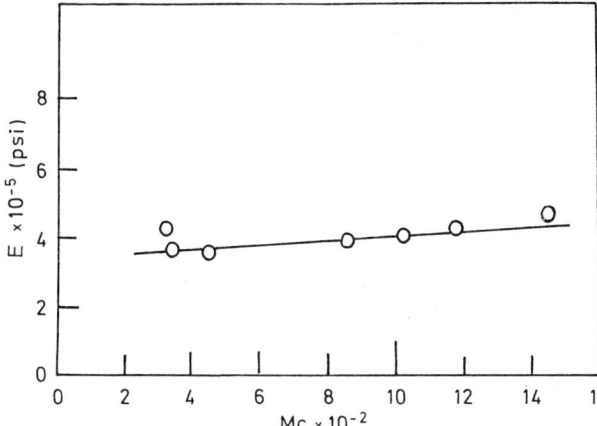

Fig. 16. Initial modulus (E) as function of Mc for DDM cured resins [53]

However, the validity of Eq. (21) for the systems montioned in chapter 3 cannot be proved. As shown in Fig. 17, hydrazide-cured resins having a maximum in $\delta^2$ also exhibit a maximum E'. Besides, the solubility parameter ($\delta^2$) computed from the values reported by Hoy [55] is 9.9 for the anhydride-cured resins and 10.6 for the DETA-cured resins, while E' of the anhydride-cured resins is higher than that of the DETA-cured resin. These results have been derived from the ten-second shear modulus [3G(10)] and flexural modulus (Table 7) [40, 46]. In the resins obtained according to Eqs. (4) to (7) (e.g., anhydride-cured resins), the network formation is not perfect and there exist many dangling chains.

The values of E' in the glassy state of the resin cured with aromatic hardeners are of the same order of magnitude as that of the resins cured with aliphatic hardeners, but the differences between the values of E' of the aromatic series are much smaller than those of the aliphatic series (Fig. 17, Table 3).

Since the modulus of elasticity in the glassy region is influenced by annealing, special care is required. When a quenched sample is annealed at a temperature a little lower than $T_g$, the modulus of elasticity is raised as shown in Table 7, and so-called enthalpy relaxation is observed as from the heat content change determined by DSC [57]. The change of the modulus of elasticity by annealing and quenching is reversible.

The tensile or bending strengths reflect the resistance of the material against fracture, and fracture is initiated by the stress concentration at defects existing in the material.

Table 7. Elasticity modulus of cured DGEBA [56]

| Curing agent | MeTHPA | DETA |
|---|---|---|
| 3G (10), dyne/cm² | | |
| Quenched | $2.68 \times 10^{10}$ | $2.45 \times 10^{10}$ |
| Annealed | $2.78 \times 10^{10}$ | $2.63 \times 10^{10}$ |
| Flexural modulus, kg/cm² | | |
| Quenched | $2.69 \times 10^4$ | $2.51 \times 10^4$ |
| Annealed | $3.08 \times 10^4$ | $2.63 \times 10^4$ |

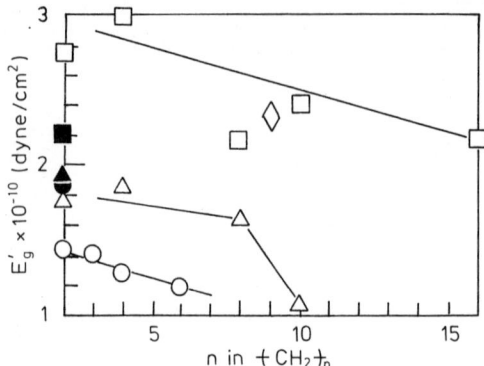

Fig. 17. Effect of the number of methylene groups (n) in the curing agent on the dynamic modulus at 20 °C.
□: aliphatic dihydrazide, △: anhydride, ○: diamine, ◇: polymercaptane, ■: aromatic dihydrazide, ▲: anhydride, ●: diamine

For many polymer materials, the following relationships between the tensile strength ($\sigma_B$), yield strength ($\sigma_E$), and modulus of elasticity holds [1, 58, 59]:

$$\sigma_B \approx E/30 \quad (22)$$

$$1/60 \leq \sigma_E/E \leq 1/30 \quad (23)$$

It has been proved that these approximate relations are applicable to many polymeric materials under various conditions including the changes of temperature, pressure, and molecular orientation [59].

The flexural strength of the cured epoxy resin decreases in the series hydrazide > anhydride > amine, which is similar to the decrease of E' in the glassy state as shown in Fig. 18. The flexural strength slightly decreases with decreasing $\varrho_{(E')}$ [37, 41, 66]. A similar tendency is observed in the resins cured with aromatic hardeners, but there is little difference among different kinds of hardeners. The dependences are stronger for resins cured with aliphatic hardeners (Table 3). However, Eq. (22) is not exactly applicable. The value of $E/\sigma_E$, is 23 in average in the resins cured with amine hardeners, 27 in those cured with anhydride hardeners, and 26 in those cured

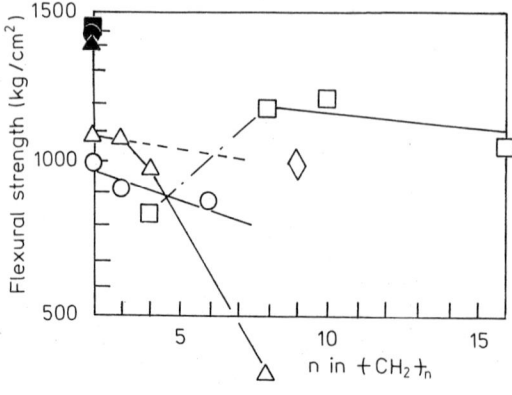

Fig. 18. Effect of the number of methylene groups (n) in the curing agent on the flexural strength.
□: aliphatic dihydrazide, △: anhydride, ○: diamine, ◇: polymercaptan, ■: aromatic dihydrazide, ▲: anhydride, ●: diamine

with hydrazide hardeners. It is necessary to further study resin-hardener systems to verify these relationships. It can be concluded that the strength of all cured resins is similar in the glassy state.

Another ultimate property which is difficult to interpret theoretically is the impact strength. For the relation between the impact strength and dynamic mechanical properties, the following equation was derived by Wada using the Maxwell model [62]:

$$\text{Impact strength (I)} = K(G'_{100} - G'_{300})^2/G'_{100} \qquad (24)$$

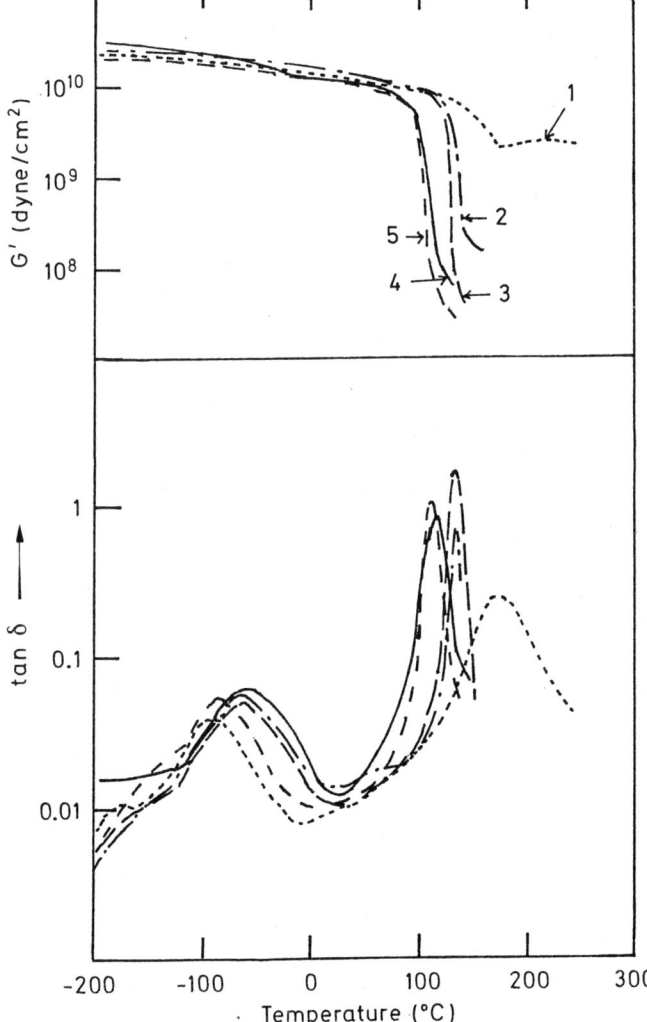

**Fig. 19.** Dynamic mechanical properties as a function of epoxy curatives [65]. Hardener; 1: PMDA, 2: HHPA, 3: MPDA, 4: DETA, 5: DMP-30

where K is a constant, $G'_{100}$ is the dynamic shear modulus at 100 K, and $G'_{300}$ is the dynamic shear modulus at 300 K.

Furthermore, the following equation was derived from the relation between I and loss modulus (tan δ):

$$I = \int_{100K}^{300K} (\tan \delta) \, dT \tag{25}$$

This equation is applicable to many polymeric materials.

On the other hand, many studies on dynamic mechanical properties at temperatures lower than room temperatures have been reported [5, 7, 63, 64]. For example, a small β transition near −50 °C has also been observed in epoxy resins. Cuddihy [65, 66] observed a β transition in resins cured with different hardeners such as DETA, MPDA, HHPA, pyromellitic dianhydride (PMDA), and tris(dimethylaminomethyl)-phenol (DMP-30) (Fig. 19). The larger the size of the β transition, the higher the impact strength (Table 8).

Sugano reports that the area of the β transition is proportional to the impact strength [67]. This result was obtained on a series of DGEBA resins cured with different amounts of methyl tetrahydrophthalic anhydride (MeTHPA) with Im as a catalyst. The area of β transition is the largest when MeTHPA/Epoxy = 0.4 (Fig. 20).

Delatyki [63] determined the modulus of resins cured with several kinds of α,ω-diamine hardeners and found that the β dispersion was due to breaking of hydrogen bonds in hydroxy ether units. May [5] and also Dammont [7], observed that the β dispersion appeared in all cured resins, and concluded that the β dispersion was associated with the relaxation of the hydroxy ether units. Arridge [64] assumed that the apparent activation energy was associated with the steric hindrance of hydroxy ether units undergoing crankshaft type rotation. According to Cuddihy [65, 66], the β dispersion was determined by the relaxation of the diether bond in Bisphenol A since β dispersion was observed even in the tertiary amine (DMP-30)-cured resin having no hydroxy ether unit (Fig. 19). Fukusawa [68, 69] assumed that the β dispersion was not associated with the relaxation of small structural units in crosslinked molecules, but with the relaxation in the whole crosslinked molecules by rotatory vibration, i.e., with the local mode relaxation. Williams [70], considering the size of the β dispersion in the resins cured with various amounts of amine hardeners considers the β dispersion to be related

**Table 8.** Impact strength and properties of cured epoxy resins [65]

| Curing agent | $T_g$ (°C) | β-transition $T_{max}$ (°C) | tan $\delta_{max}$ | Izod impact[b] (ft-lb/in) |
|---|---|---|---|---|
| DETA | 112 | −68 | .0626 | .3–.4 |
| DMP-30 | 108 | −85 | .0525 | ? |
| MPDA | 130 | −64 | .0489 | .207 |
| HHPA | 134 | −65 | .0563 | .3 |
| PMDA | 148 | −95 | .0415 | .1 |

[a] Estimated using Tobolsky's rule;
[b] Reference [38]

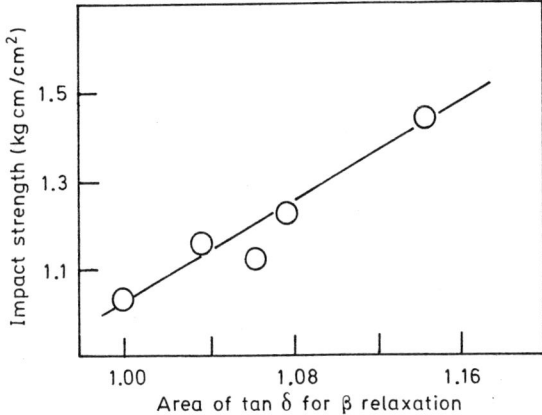

**Fig. 20.** Correlation between the area of β relaxation and the impact strength of MeTHPA cured resins [67]

to the hydroxy ether unit and bisphenol unit as well. In the epoxy-isocyanate resin mentioned in Chapter 4.2, the β dispersion is located at the same temperature and is not related to the structure of the epoxy prepolymers (Fig. 15). Since these resins consist almost entirely of cyclic structures and have no hydroxy ether unit, it is believed that the β dispersion is associated with the local mode relaxation. Therefore, the origin of the β dispersion is not clear and more studies are necessary.

Since such strength of break at very high strain speeds (as e.g., in the impact) is equal to the strength at low temperatures, it is assumed that the β dispersion may absorb the breaking energy and somewhat influence the impact strength.

# 6 Conclusions

Cured epoxy resins are often used as structural materials and their performance is determined by their mechanical properties in the glassy state. However, as mentioned in Chapter 5, the mechanical properties of unfilled resins in the glassy state are almost equal and do not depend much on the chemical structure and degree of crosslinking.

The value of $T_g$ is important, since rigidity decreases in the glass transition region. $T_g$ is closely related to the structure and crosslinking density of the cured resin. The structure of the cured resin can be derived from the initial starting materials, the epoxy prepolymer and hardener, and reaction conditions. However, the structures of many cured resins is still unclear which prevents to establish the structure-mechanical properties relationships. Further studies are needed. Furthermore, the commercial epoxy formulations may contain several components and also diluents, plasticizers, liquid rubbers, etc. which makes a prediction of $T_g$ and mechanical properties even more difficult.

Finally, it should be emphasized that the mechanical properties discussed here have been determined on resins cured as perfectly as possible. However, the proof of perfect curing is still difficult. Therefore, to establish relationships between the structure of the cured resins and their mechanical properties, it will be necessary to work out methods for determining the degree of cure and procedures to prepare perfectly cured resins.

# 7 Appendix

| | |
|---|---|
| $E'$: | dynamic Young's modulus |
| $T_g$: | glass transition temperature |
| $\varrho$: | crosslinking density |
| $\varrho_{(E')}$: | value of $\varrho$ obtained by assuming that $\varphi$ is equal to 1 |
| $M_c$: | molecular weight between crosslinks |
| d | density |
| R: | gas constant |
| T: | absolute temperature |
| $\varphi$: | front factor |
| tan $\delta$: | loss tangent |
| $G'$: | shear modulus |
| $J'$: | shear compliance |
| E: | Young's modulus |
| B: | bulk modulus of elasticity |
| $\delta$: | solubility parameter |
| $\sigma_B$: | tensil strength |
| $\sigma_E$: | tensile yield strength |

| | |
|---|---|
| BDA: | 1,4-butylenediamine |
| MPDA: | *m*-phenylenediamine |
| PA: | phthalic acid anhydride |
| MNA: | methylnadic acid anhydride |
| HHPA: | hexahydrophthalic acid anhydride |
| Me-HHPA: | methyl-hexahydrophthalic acid anhydride |
| Suc A: | succinic acid anhydride |
| MA: | maleic acid anhydride |
| Me-THPA: | methyl-tetrahydrophthalic acid anhydride |
| DMP-30: | tris(dimethylaminomethyl)phnol |

# 8 References

1. Nielsen, L. E.: Mechanical Properties of Polymer and Composite, Chap. 4 Marcel Dekker, New York 1975
2. Kaelble, D. H.: SPE J. *15*, 1071 (1959)
3. Kline, D. E.: J. Polym. Sci. *47*, 237 (1960)
4. Kline, D. E., Sauer, J. A.: SPE Trans. *2*, 21 (1962)
5. May, C. A., Weir, F. E.: SPE Trans. *2*, 207 (1962)
6. Kwei, T. K.: J. Polym. Sci., A-2, *4*, 943 (1966)
7. Dammont, F. R., Kwei, T. K.: ibid. A-2, *5*, 764 (1967)
8. Murayama, T., Bell, J. P.: ibid. A-2, *8*, 437 (1970)
9. Smith, I. T.: Polymer *2*, 92 (1961)
10. Kakurai, T., Noguchi, T.: Kogyo Kagaku Zashi *63*, 294 (1960)

11. Tanaka, Y., Mika, T. F.: Epoxy Resins Chemistry and Technology, p. 135, Marcel Dekker, New York 1973
12. King, J. J., Bell, J. P.: Epoxy Resin Chemistry, ACS Symp. Vol. 114, ACS (1979), p. 225
13. Dusek, K., Bleha, M., Lunak, S.: J. Polym. Sci., Polyr. Chem. Ed. *15*, 2393 (1977)
14. Fish, W., Hofmann, W.: ibid. *12*, 497 (1954)
15. Tanaka, Y., Kakiuchi, H.: ibid. A-2, 3405 (1964)
16. Fischer, R. F.: ibid. *44*, 155 (1960)
17. Matejka, L., Lovy, J., Pokorny, S., Bouchal, K., Dusek, K.: ibid. Poly. Chem. Ed. *21*, 2873 (1983)
18. Narracott: Brit. Plastics *26*, 120 (1953)
19. Saito, K., Kamon, T., Miwa, Y., Saeki, K.: Shikizai *51*, 271 (1978)
20. Farkas, A., Strohm, P. F.: J. Appl. Polymer Sci. *12*, 159 (1968)
21. Kamon, T., Saito, K., Miwa, Y., Saeki, K.: Shikizai *49*, 82 (1976)
22. Harris, J. J., Temine, S. C.: J. Appl. Polymer Sci. *10*, 523 (1966)
23. Kamon, T., Saito, K., Miwa, Y., Saeki, K.: Nippon Kagaku Kaishi 1958 (1972)
24. Crivello, L. V., Lam, J. H. W.: Epoxy Resin Chemistry, ACS Symp. Vol. 114, ACS (1979), p. 1; Watt, W. R., ibid. p. 17
25. Izumo, T.: Shikizai *49*, 538 (1976)
26. Kamon, T., Saito, K.: Konunshi Ronbubshu *37*, 765 (1980)
27. Kamon, T., Saito, K., Miwa, Y., Saeki, K.: ibid. *31*, 119 (1974)
28. Tobolsky, A. V.: Properties and Structure of Polymer John Wiley, New York 1960
29. Nielsen, L. E.: J. Macromol. Sci., Revs. Macromol. Chem. *C3*, 69 (1969)
30. Fox, T. G., Loshaek, S.: J. Polym., Sci. *15*, 371 (1955)
31. Diamant, Y., Welner, S., Katz, D.: Polymer *11*, 498 (1970)
32. Shibayama, K.: Kobnshi Kagaku *18*, 183 (1961)
33. Kamon, T., Saito, K., Miwa, Y., Saeki, K.: ibid. *30*, 279 (1973)
34. Kamon, T.: Kobunshi Ronbunshu *36*, 597 (1979)
35. Katz, D., Tobolsky, A. V.: Polymer *4*, 417 (1963)
36. Kamon, T., Saito, K.: Kobunshi Ronbunshu *37*, 765 (1980)
37. Kamon, T., Saito, K.: Shikizai *54*, 416 (1981)
38. Lee, H., Neville, K.: Handbook of Epoxy Resins, McGraw-Hill, New York 1967
39. Miyamoto, A., Shibayama, S.: Kobunshi kagaku *29*, 325 (1972)
40. Batzer, H., Lohse, F., Schmid, R.: Angew. Makromol. Chem. *29/30*, 349 (1973)
41. Kamon, T., Saito, K., Takii, K.: Netsukokasei jushi *3*, 127 (1982)
42. Bruins, P. F.: Epoxy Resin Technology, pp. 166-8, Interscience 1968
43. Iwakura, Y., Okada, H., Yoshida, Y.: Makromol. Chem. *98*, 7 (1966)
44. Kamon, T., Saito, K.: Netsukokasei jushi *2*, 140 (1981)
45. Kamon, T., Saito, K.: ibid. *5*, 91 (1984)
46. Danehy, J. P., Noel, C. J.: Amer. Chem. Soc. *82*, 2511 (1960)
47. Kamon, T., Saito, K.: Kobunshi Ronbunshu *34*, 537 (1977)
48. Saunders, T. F., Levy, M. F., Serino, J. F.: J. Polym. Sci., A-1, *5*, 1609 (1967)
49. Sandler, S. R., Berg, F., Kitazawa, G.: J. Appl. Polym. Sci. *9*, 1994 (1965)
50. Bald, G., Kretzmar, K., Markert, H., Wimmer, M.: Angew. Makromol. Chem. *44*, 151 (1975)
51. Kinjo, N., Numata, D., Koyama, T., Katsuya, Y.: Polymer J. *14*, 505 (1982)
52. Catsiff, E. H., Bee, H. B., DiPrima, J. F., Seltzer, R.: ACS Polymer Div., preprint, p. 111 (1981)
53. Bell, J. P.: J. Appl. Polym. Sci. *14*, 1901 (1970)
54. Holliday, L., White, J. W.: Pure Appl. Chem. *26*, 545 (1971)
55. Hoy, K. L.: J. Paint Technol. *42*, 76 (1970)
56. Hata, N.: Netsukokasei jushi *2*, 29 (1981)
57. Kreibich, U., Schmid, R.: J. Polym. Sci., Symp. No. 53, 177 (1975)
58. Buchdahl, R.: J. Polym. Sci. *28*, 239 (1958)
59. Brown, N.: Mater. Sci. Eng. *8*, 65 (1971)
60. Kamon, T.: Nippon Sechaku Kyokaishi *15*, 107 (1979)
61. Shimbo, M., Iwakoshi, M., Ochi, K.: Nippon Sechaku Kyokaishi *10*, 161 (1974)
62. Wada, Y., Kasahara, T.: J. Appl. Polym. Sci. *11*, 1661 (1967)
63. Delatyki, O., Shaw, J. C., Williams, J. G.: J. Polym. Sci. A-2, *7*, 753 (1969)
64. Arridge, R. G. C., Speake, J. H.: Polymer *13*, 443 (1972)

65. Cuddihy, E. F., Moacanin, J.: 155th ACS meeting-Org. Coating & Plastic Chemistry preprint *28*(7), 449 (1968)
66. Cuddihy, E. F., Moacanin, J.: J. Polym. Sci. A-2, *8*, 1627 (1970)
67. Sugano, T., Miyamoto, T., Shibayama, K.: 21th Thermoset resin Conf. in Japan, preprint, p. 41A (1971)
68. Fukusawa, Y., Wada, E.: Kobunshi Ronbunshu *31*, 186 (1974)
69. Fukasawa, Y., Wada, E.: ibid. *32*, 518 (1975)
70. Williams, J. G.: J. Appl. Polymer, Sci. *23*, 7433 (1979)

Editor: K. Dušek
Received: October 25, 1985

# Author Index Volumes 1–80

*Allegra, G.* and *Bassi, I. W.:* Isomorphism in Synthetic Macromolecular Systems. Vol. 6, pp. 549–574.
*Andrade, J. D., Hlady, V.:* Protein Adsorption and Materials Biocompability: A. Tutorial Review and Suggested Hypothesis. Vol. 79, pp. 1–63.
*Andrews, E. H.:* Molecular Fracture in Polymers. Vol. 27, pp. 1–66.
*Anufrieva, E. V.* and *Gotlib, Yu. Ya.:* Investigation of Polymers in Solution by Polarized Luminescence. Vol. 40, pp. 1–68.
*Apicella, A.* and *Nicolais, L.:* Effect of Water on the Properties of Epoxy Matrix and Composite. Vol. 72, pp. 69–78.
*Apicella, A., Nicolais, L.* and *de Cataldis, C.:* Characterization of the Morphological Fine Structure of Commercial Thermosetting Resins Through Hygrothermal Experiments. Vol. 66, pp. 189–208.
*Argon, A. S., Cohen, R. E., Gebizlioglu, O. S.* and *Schwier, C.:* Crazing in Block Copolymers and Blends. Vol. 52/53, pp. 275–334.
*Aronhime, M. T., Gillham, J. K.:* Time-Temperature Transformation (TTT) Cure Diagram of Thermosetting Polymeric Systems. Vol. 78, pp. 81–112.
*Arridge, R. C.* and *Barham, P. J.:* Polymer Elasticity. Discrete and Continuum Models. Vol. 46, pp. 67–117.
*Aseeva, R. M., Zaikov, G. E.:* Flammability of Polymeric Materials. Vol. 70, pp. 171–230.
*Ayrey, G.:* The Use of Isotopes in Polymer Analysis. Vol. 6, pp. 128–148.


*Bässler, H.:* Photopolymerization of Diacetylenes. Vol. 63, pp. 1–48.
*Baldwin, R. L.:* Sedimentation of High Polymers. Vol. 1, pp. 451–511.
*Balta-Calleja, F. J.:* Microhardness Relating to Crystalline Polymers. Vol. 66, pp. 117–148.
*Barton, J. M.:* The Application of Differential Scanning Calorimetry (DSC) to the Study of Epoxy Resins Curing Reactions. Vol. 72, pp. 111–154.
*Basedow, A. M.* and *Ebert, K.:* Ultrasonic Degradation of Polymers in Solution. Vol. 22, pp. 83–148.
*Batz, H.-G.:* Polymeric Drugs. Vol. 23, pp. 25–53.
*Bell, J. P.* see *Schmidt, R. G.:* Vol. 75, pp. 33–72.
*Bekturov, E. A.* and *Bimendina, L. A.:* Interpolymer Complexes. Vol. 41, pp. 99–147.
*Bergsma, F.* and *Kruissink, Ch. A.:* Ion-Exchange Membranes. Vol. 2, pp. 307–362.
*Berlin, Al. Al., Volfson, S. A.,* and *Enikolopian, N. S.:* Kinetics of Polymerization Processes. Vol. 38, pp. 89–140.
*Berry, G. C.* and *Fox, T. G.:* The Viscosity of Polymers and Their Concentrated Solutions. Vol. 5, pp. 261–357.
*Bevington, J. C.:* Isotopic Methods in Polymer Chemistry. Vol. 2, pp. 1–17.
*Bhuiyan, A. L.:* Some Problems Encountered with Degradation Mechanisms of Addition Polymers. Vol. 47, pp. 1–65.
*Bird, R. B., Warner, Jr., H. R.,* and *Evans, D. C.:* Kinetik Theory and Rheology of Dumbbell Suspensions with Brownian Motion. Vol. 8, pp. 1–90.
*Biswas, M.* and *Maity, C.:* Molecular Sieves as Polymerization Catalysts. Vol. 31, pp. 47–88.
*Biswas, M., Packirisamy, S.:* Synthetic Ion-Exchange Resins. Vol. 70, pp. 71–118.
*Block, H.:* The Nature and Application of Electrical Phenomena in Polymers. Vol. 33, pp. 93–167.

*Bodor, G.:* X-ray Line Shape Analysis. A. Means for the Characterization of Crystalline Polymers. Vol. 67, pp. 165–194.
*Böhm, L. L., Chmeliř, M., Löhr, G., Schmitt, B. J.* and *Schulz, G. V.:* Zustände und Reaktionen des Carbanions bei der anionischen Polymerisation des Styrols. Vol. 9, pp. 1–45.
*Bovey, F. A.* and *Tiers, G. V. D.:* The High Resolution Nuclear Magnetic Resonance Spectroscopy of Polymers. Vol. 3, pp. 139–195.
*Braun, J.-M.* and *Guillet, J. E.:* Study of Polymers by Inverse Gas Chromatography. Vol. 21, pp. 107–145.
*Breitenbach, J. W., Olaj, O. F.* und *Sommer, F.:* Polymerisationsanregung durch Elektrolyse. Vol. 9, pp. 47–227.
*Bresler, S. E.* and *Kazbekov, E. N.:* Macroradical Reactivity Studied by Electron Spin Resonance. Vol. 3, pp. 688–711.
*Bucknall, C. B.:* Fracture and Failure of Multiphase Polymers and Polymer Composites. Vol. 27, pp. 121–148.
*Burchard, W.:* Static and Dynamic Light Scattering from Branched Polymers and Biopolymers. Vol. 48, pp. 1–124.
*Bywater, S.:* Polymerization Initiated by Lithium and Its Compounds. Vol. 4, pp. 66–110.
*Bywater, S.:* Preparation and Properties of Star-branched Polymers. Vol. 30, pp. 89–116.


*Candau, S., Bastide, J.* und *Delsanti, M.:* Structural. Elastic and Dynamic Properties of Swollen Polymer Networks. Vol. 44, pp. 27–72.
*Carrick, W. L.:* The Mechanism of Olefin Polymerization by Ziegler-Natta Catalysts. Vol. 12, pp. 65–86.
*Casale, A.* and *Porter, R. S.:* Mechanical Synthesis of Block and Graft Copolymers. Vol. 17, pp. 1–71.
*Cerf, R.:* La dynamique des solutions de macromolecules dans un champ de vitresses. Vol. 1, pp. 382–450.
*Cesca, S., Priola, A.* and *Bruzzone, M.:* Synthesis and Modification of Polymers Containing a System of Conjugated Double Bonds. Vol. 32, pp. 1–67.
*Chiellini, E., Solaro, R., Galli, G.* and *Ledwith, A.:* Optically Active Synthetic Polymers Containing Pendant Carbazolyl Groups. Vol. 62, pp. 143–170.
*Cicchetti, O.:* Mechanisms of Oxidative Photodegradation and of UV Stabilization of Polyolefins. Vol. 7, pp. 70–112.
*Clark, D. T.:* ESCA Applied to Polymers. Vol. 24, pp. 125–188.
*Colemann, Jr., L. E.* and *Meinhardt, N. A.:* Polymerization Reactions of Vinyl Ketones. Vol. 1, pp. 159–179.
*Comper, W. D.* and *Preston, B. N.:* Rapid Polymer Transport in Concentrated Solutions. Vol. 55, pp. 105–152.
*Corner, T.:* Free Radical Polymerization — The Synthesis of Graft Copolymers. Vol. 62, pp. 95–142.
*Crescenzi, V.:* Some Recent Studies of Polyelectrolyte Solutions. Vol. 5, pp. 358–386.
*Crivello, J. V.:* Cationic Polymerization — Iodonium and Sulfonium Salt Photoinitiators, Vol. 62, pp. 1–48.


*Dave, R.* see *Kardos, J. L.:* Vol. 80, pp. 101–123.
*Davydov, B. E.* and *Krentsel, B. A.:* Progress in the Chemistry of Polyconjugated Systems. Vol. 25, pp. 1–46.
*Dettenmaier, M.:* Intrinsic Crazes in Polycarbonate Phenomenology and Molecular Interpretation of a New Phenomenon. Vol. 52/53, pp. 57–104.
*Dobb, M. G.* and *McIntyre, J. E.:* Properties and Applications of Liquid-Crystalline Main-Chain Polymers. Vol. 60/61, pp. 61–98.
*Döll, W.:* Optical Interference Measurements and Fracture Mechanics Analysis of Crack Tip Craze Zones. Vol. 52/53, pp. 105–168.
*Doi, Y.* see *Keii, T.:* Vol. 73/74, pp. 201–248.
*Dole, M.:* Calorimetric Studies of States and Transitions in Solid High Polymers. Vol. 2, pp. 221–274.

*Donnet, J. B., Vidal, A.:* Carbon Black-Surface Properties and Interactions with Elastomers. Vol. 76, pp. 103–128.
*Dorn, K., Hupfer, B.,* and *Ringsdorf, H.:* Polymeric Monolayers and Liposomes as Models for Biomembranes How to Bridge the Gap Between Polymer Science and Membrane Biology? Vol. 64, pp. 1–54.
*Dreyfuss, P.* and *Dreyfuss, M. P.:* Polytetrahydrofuran. Vol. 4, pp. 528–590.
*Drobnik, J.* and *Rypáček, F.:* Soluble Synthetic Polymers in Biological Systems. Vol. 57, pp. 1–50.
*Dröscher, M.:* Solid State Extrusion of Semicrystalline Copolymers. Vol. 47, pp. 120–138.
*Duduković, M. P.* see Kardos, J. L.: Vol. 80, pp. 101–123.
*Drzal, L. T.:* The Interphase in Epoxy Composites. Vol. 75, pp. 1–32.
*Dušek, K.:* Network Formation in Curing of Epoxy Resins. Vol. 78, pp. 1–58.
*Dušek, K.* and *Prins, W.:* Structure and Elasticity of Non-Crystalline Polymer Networks. Vol. 6, pp. 1–102.
*Duncan, R.* and *Kopeček, J.:* Soluble Synthetic Polymers as Potential Drug Carriers. Vol. 57, pp. 51–101.

*Eastham, A. M.:* Some Aspects of the Polymerization of Cyclic Ethers. Vol. 2, pp. 18–50.
*Ehrlich, P.* and *Mortimer, G. A.:* Fundamentals of the Free-Radical Polymerization of Ethylene. Vol. 7, pp. 386–448.
*Eisenberg, A.:* Ionic Forces in Polymers. Vol. 5, pp. 59–112.
*Eiss, N. S. Jr.* see Yorkgitis, E. M. Vol. 72, pp. 79–110.
*Elias, H.-G., Bareiss, R.* und *Watterson, J. G.:* Mittelwerte des Molekulargewichts und anderer Eigenschaften. Vol. 11, pp. 111–204.
*Elsner, G., Riekel, Ch.* and *Zachmann, H. G.:* Synchrotron Radiation Physics. Vol. 67, pp. 1–58.
*Elyashevich, G. K.:* Thermodynamics and Kinetics of Orientational Crystallization of Flexible-Chain Polymers. Vol. 43, pp. 207–246.
*Enkelmann, V.:* Structural Aspects of the Topochemical Polymerization of Diacetylenes. Vol. 63, pp. 91–136.
*Entelis, S. G., Evreinov, V. V., Gorshkov, A. V.:* Functionally and Molecular Weight Distribution of Telchelic Polymers. Vol. 76, pp. 129–175.
*Evreinov, V. V.* see Entelis, S. G. Vol. 76, pp. 129–175.

*Ferruti, P.* and *Barbucci, R.:* Linear Amino Polymers: Synthesis, Protonation and Complex Formation. Vol. 58, pp. 55–92.
*Finkelmann, H.* and *Rehage, G.:* Liquid Crystal Side-Chain Polymers. Vol. 60/61, pp. 99–172.
*Fischer, H.:* Freie Radikale während der Polymerisation, nachgewiesen und identifiziert durch Elektronenspinresonanz. Vol. 5, pp. 463–530.
*Flory, P. J.:* Molecular Theory of Liquid Crystals. Vol. 59, pp. 1–36.
*Ford, W. T.* and *Tomoi, M.:* Polymer-Supported Phase Transfer Catalysts Reaction Mechanisms. Vol. 55, pp. 49–104.
*Fradet, A.* and *Maréchal, E.:* Kinetics and Mechanisms of Polyesterifications. I. Reactions of Diols with Diacids. Vol. 43, pp. 51–144.
*Franz, G.:* Polysaccharides in Pharmacy. Vol. 76, pp. 1–30.
*Friedrich, K.:* Crazes and Shear Bands in Semi-Crystalline Thermoplastics. Vol. 52/53, pp. 225–274.
*Fujita, H.:* Diffusion in Polymer-Diluent Systems. Vol. 3, pp. 1–47.
*Funke, W.:* Über die Strukturaufklärung vernetzter Makromoleküle, insbesondere vernetzter Polyesterharze, mit chemischen Methoden. Vol. 4, pp. 157–235.
*Furukawa, H.* see Kamon, T.: Vol. 80, pp. 173–202.

*Gal'braikh, L. S.* and *Rigovin, Z. A.:* Chemical Transformation of Cellulose. Vol. 14, pp. 87–130.
*Galli, G.* see Chiellini, E. Vol. 62, pp. 143–170.
*Gallot, B. R. M.:* Preparation and Study of Block Copolymers with Ordered Structures, Vol. 29, pp. 85–156.
*Gandini, A.:* The Behaviour of Furan Derivatives in Polymerization Reactions. Vol. 25, pp. 47–96.

*Gandini, A.* and *Cheradame, H.:* Cationic Polymerization. Initiation with Alkenyl Monomers. Vol. 34/35, pp. 1–289.
*Geckeler, K., Pillai, V. N. R.,* and *Mutter, M.:* Applications of Soluble Polymeric Supports. Vol. 39, pp. 65–94.
*Gerrens, H.:* Kinetik der Emulsionspolymerisation. Vol. 1, pp. 234–328.
*Ghiggino, K. P., Roberts, A. J.* and *Phillips, D.:* Time-Resolved Fluorescence Techniques in Polymer and Biopolymer Studies. Vol. 40, pp. 69–167.
*Gillham, J. K.* see Aronhime, M. T.: Vol. 78, pp. 81–112.
*Glöckner, G.:* Analysis of Compositional and Structural Heterogeneitis of Polymer by Non-Exclusion HPLC. Vol. 79, pp. 159–214.
*Godovsky, Y. K.:* Thermomechanics of Polymers. Vol. 76, pp. 31–102.
*Goethals, E. J.:* The Formation of Cyclic Oligomers in the Cationic Polymerization of Heterocycles. Vol. 23, pp. 103–130.
*Gorshkov, A. V.* see Entelis, S. G. Vol. 76, 129–175.
*Graessley, W. W.:* The Etanglement Concept in Polymer Rheology. Vol. 16, pp. 1–179.
*Graessley, W. W.:* Entagled Linear, Branched and Network Polymer Systems. Molecular Theories. Vol. 47, pp. 67–117.
*Grebowicz, J.* see Wunderlich, B. Vol. 60/61, pp. 1–60.
*Greschner, G. S.:* Phase Distribution Chromatography. Possibilities and Limitations. Vol. 73/74, pp. 1–62.

*Hagihara, N., Sonogashira, K.* and *Takahashi, S.:* Linear Polymers Containing Transition Metals in the Main Chain. Vol. 41, pp. 149–179.
*Hasegawa, M.:* Four-Center Photopolymerization in the Crystalline State. Vol. 42, pp. 1–49.
*Hatano, M.:* Induced Circular Dichroism in Biopolymer-Dye System. Vol. 77, pp. 1–121.
*Hay, A. S.:* Aromatic Polyethers. Vol. 4, pp. 496–527.
*Hayakawa, R.* and *Wada, Y.:* Piezoelectricity and Related Properties. of Polymer Films. Vol. 11, pp. 1–55.
*Heidemann, E.* and *Roth, W.:* Synthesis and Investigation of Collagen Model Peptides. Vol. 43, pp. 145–205.
*Heitz, W.:* Polymeric Reagents. Polymer Design, Scope, and Limitations. Vol. 23, pp. 1–23.
*Helfferich, F.:* Ionenaustausch. Vol. 1, pp. 329–381.
*Hendra, P. J.:* Laser-Raman Spectra of Polymers. Vol. 6, pp. 151–169.
*Hendrix, J.:* Position Sensitive "X-ray Detectors". Vol. 67, pp. 59–98.
*Henrici-Olivé, G.* and *Olivé, S.:* Oligomerization of Ethylene with Soluble Transition-Metal Catalysts. pp. 496–577.
*Henrici-Olivé, G.* und *Olivé, S.:* Koordinative Polymerisation an löslichen Übergangsmetall-Katalysatoren. Vol. 6, pp. 421–472.
*Henrici-Olivé, G.* and *Olivé, S.:* Oligomerization of Ethylene with Soluble Transition-Metal Catalysts. Vol. 15, pp. 1–30.
*Henrici-Olivé, G.* and *Olivé, S.:* Molecular Interactions and Macroscopic Properties of Polyacrylonitrile and Model Substances. Vol. 32, pp. 123–152.
*Henrici-Olivé, G.* and *Olivé, S.:* The Chemistry of Carbon Fiber Formation from Polyacrylonitrile. Vol. 51, pp. 1–60.
*Hermans, Jr., J., Lohr, D.* and *Ferro, D.:* Treatment of the Folding and Unfolding of Protein Molecules in Solution According to a Lattic Model. Vol. 9, pp. 229–283.
*Higashimura, T.* and *Sawamoto, M.:* Living Polymerization and Selective Dimerization: Two Extremes of the Polymer Synthesis by Cationic Polymerization. Vol. 62, pp. 49–94.
*Hlady, V.* see Andrade, J. D.: Vol. 79, pp. 1–63.
*Hoffman, A. S.:* Ionizing Radiation and Gas Plasma (or Glow) Discharge Treatments for Preparation of Novel Polymeric Biomaterials. Vol. 57, pp. 141–157.
*Holzmüller, W.:* Molecular Mobility, Deformation and Relaxation Processes in Polymers. Vol. 26, pp. 1–62.
*Hutchison, J.* and *Ledwith, A.:* Photoinitiation of Vinyl Polymerization by Aromatic Carbonyl Compounds. Vol. 14, pp. 49–86.

*Iizuka, E.:* Properties of Liquid Crystals of Polypeptides: with Stress on the Electromagnetic Orientation. Vol. 20, pp. 79–107.
*Ikada, Y.:* Characterization of Graft Copolymers. Vol. 29, pp. 47–84.
*Ikada, Y.:* Blood-Compatible Polymers. Vol. 57, pp. 103–140.
*Imanishi, Y.:* Synthese, Conformation, and Reactions of Cyclic Peptides. Vol. 20, pp. 1–77.
*Inagaki, H.:* Polymer Separation and Characterization by Thin-Layer Chromatography. Vol. 24, pp. 189–237.
*Inoue, S.:* Asymmetric Reactions of Synthetic Polypeptides. Vol. 21, pp. 77–106.
*Ise, N.:* Polymerizations under an Electric Field. Vol. 6, pp. 347–376.
*Ise, N.:* The Mean Activity Coefficient of Polyelectrolytes in Aqueous Solutions and Its Related Properties. Vol. 7, pp. 536–593.
*Isihara, A.:* Intramolecular Statistics of a Flexible Chain Molecule. Vol. 7, pp. 449–476.
*Isihara, A.:* Irreversible Processes in Solutions of Chain Polymers. Vol. 5, pp. 531–567.
*Isihara, A.* and *Guth, E.:* Theory of Dilute Macromolecular Solutions. Vol. 5, pp. 233–260.
*Iwatsuki, S.:* Polymerization of Quinodimethane Compounds. Vol. 58, pp. 93–120.

*Janeschitz-Kriegl, H.:* Flow Birefrigence of Elastico-Viscous Polymer Systems. Vol. 6, pp. 170–318.
*Jenkins, R.* and *Porter, R. S.:* Unpertubed Dimensions of Stereoregular Polymers. Vol. 36, pp. 1–20.
*Jenngins, B. R.:* Electro-Optic Methods for Characterizing Macromolecules in Dilute Solution. Vol. 22, pp. 61–81.
*Johnston, D. S.:* Macrozwitterion Polymerization. Vol. 42, pp. 51–106.

*Kamachi, M.:* Influence of Solvent on Free Radical Polymerization of Vinyl Compounds. Vol. 38, pp. 55–87.
*Kamon, T., Furukawa, H.:* Curing Mechanisms and Mechanical Properties of Cured Epoxy Resins. Vol. 80, pp. 173–202.
*Kaneko, M.* and *Yamada, A.:* Solar Energy Conversion by Functional Polymers. Vol. 55, pp. 1–48.
*Kardos, J. L., Dudukovič, M. P., Dave, R.:* Void Growth and Resin Transport During Processing of Thermosetting — Matrix Composites. Vol. 80, pp. 101–123.
*Kawabata, S.* and *Kawai, H.:* Strain Energy Density Functions of Rubber Vulcanizates from Biaxial Extension. Vol. 24, pp. 89–124.
*Keii, T., Doi, Y.:* Synthesis of "Living" Polyolefins with Soluble Ziegler-Natta Catalysts and Application to Block Copolymerization. Vol. 73/74, pp. 201–248.
*Kelley, F. N.* see *LeMay, J. D.:* Vol. 78, pp. 113–148.
*Kennedy, J. P.* and *Chou, T.:* Poly(isobutylene-co-β-Pinene): A New Sulfur Vulcanizable, Ozone Resistant Elastomer by Cationic Isomerization Copolymerization. Vol. 21, pp. 1–39.
*Kennedy, J. P.* and *Delvaux, J. M.:* Synthesis, Characterization and Morphology of Poly(butadiene-g-Styrene). Vol. 38, pp. 141–163.
*Kennedy, J. P.* and *Gillham, J. K.:* Cationic Polymerization of Olefins with Alkylaluminium Initiators. Vol. 10, pp. 1–33.
*Kennedy, J. P.* and *Johnston, J. E.:* The Cationic Isomerization Polymerization of 3-Methyl-1-butene and 4-Methyl-1-pentene. Vol. 19, pp. 57–95.
*Kennedy, J. P.* and *Langer, Jr., A. W.:* Recent Advances in Cationic Polymerization. Vol. 3, pp. 508–580.
*Kennedy, J. P.* and *Otsu, T.:* Polymerization with Isomerization of Monomer Preceding Propagation. Vol. 7, pp. 369–385.
*Kennedy, J. P.* and *Rengachary, S.:* Correlation Between Cationic Model and Polymerization Reactions of Olefins. Vol. 14, pp. 1–48.
*Kennedy, J. P.* and *Trivedi, P. D.:* Cationic Olefin Polymerization Using Alkyl Halide — Alkyl-Aluminium Initiator Systems. I. Reactivity Studies. II. Molecular Weight Studies. Vol. 28, pp. 83–151.
*Kennedy, J. P., Chang, V. S. C.* and *Guyot, A.:* Carbocationic Synthesis and Characterization of Polyolefins with Si-H and Si-Cl Head Groups. Vol. 43, pp. 1–50.
*Khoklov, A. R.* and *Grosberg, A. Yu.:* Statistical Theory of Polymeric Lyotropic Liquid Crystals. Vol. 41, pp. 53–97.

*Kinloch, A. J.:* Mechanics and Mechanisms of Fracture of Thermosetting Epoxy Polymers. Vol. 72, pp. 45–68.
*Kissin, Yu. V.:* Structures of Copolymers of High Olefins. Vol. 15, pp. 91–155.
*Kitagawa, T.* and *Miyazawa, T.:* Neutron Scattering and Normal Vibrations of Polymers. Vol. 9, pp. 335–414.
*Kitamaru, R.* and *Horii, F.:* NMR Approach to the Phase Structure of Linear Polyethylene. Vol. 26, pp. 139–180.
*Klosinski, P., Penczek, S.:* Teichoic Acids and Their Models: Membrane Biopolymers with Polyphosphate Backbones. Synthesis, Structure and Properties. Vol. 79, pp. 139–157.
*Knappe, W.:* Wärmeleitung in Polymeren. Vol. 7, pp. 477–535.
*Koenik, J. L.* see *Mertzel, E.* Vol. 75, pp. 73–112.
*Koenig, J. L.:* Fourier Transforms Infrared Spectroscopy of Polymers, Vol. 54, pp. 87–154.
*Kolařík, J.:* Secondary Relaxations in Glassy Polymers: Hydrophilic Polymethacrylates and Polyacrylates: Vol. 46, pp. 119–161.
*Kong, E. S.-W.:* Physical Aging in Epoxy Matrices and Composites. Vol. 80, pp. 125–171.
*Koningsveld, R.:* Preparative and Analytical Aspects of Polymer Fractionation. Vol. 7.
*Kovacs, A. J.:* Transition vitreuse dans les polymers amorphes. Etude phénoménologique. Vol. 3, pp. 394–507.
*Krässig, H. A.:* Graft Co-Polymerization of Cellulose and Its Derivatives. Vol. 4, pp. 111–156.
*Kramer, E. J.:* Microscopic and Molecular Fundamentals of Crazing. Vol. 52/53, pp. 1–56.
*Kraus, G.:* Reinforcement of Elastomers by Carbon Black. Vol. 8, pp. 155–237.
*Kreutz, W.* and *Welte, W.:* A General Theory for the Evaluation of X-Ray Diagrams of Biomembranes and Other Lamellar Systems. Vol. 30, pp. 161–225.
*Krimm, S.:* Infrared Spectra of High Polymers. Vol. 2, pp. 51–72.
*Kuhn, W., Ramel, A., Walters, D. H. Ebner, G.* and *Kuhn, H. J.:* The Production of Mechanical Energy from Different Forms of Chemical Energy with Homogeneous and Cross-Striated High Polymer Systems. Vol. 1, pp. 540–592.
*Kunitake, T.* and *Okahata, Y.:* Catalytic Hydrolysis by Synthetic Polymers. Vol. 20, pp. 159–221.
*Kurata, M.* and *Stockmayer, W. H.:* Intrinsic Viscosities and Unperturbed Dimensions of Long Chain Molecules. Vol. 3, pp. 196–312.

*Ledwith, A.* and *Sherrington, D. C.:* Stable Organic Cation Salts: Ion Pair Equilibria and Use in Cationic Polymerization. Vol. 19, pp. 1–56.
*Ledwith, A.* see Chiellini, E. Vol. 62, pp. 143–170.
*Lee, C.-D. S.* and *Daly, W. H.:* Mercaptan-Containing Polymers. Vol. 15, pp. 61–90.
*LeMay, J. D., Kelley, F. N.:* Structure and Ultimate Properties of Epoxy Resins. Vol. 78, pp. 113–148.
*Lindberg, J. J.* and *Hortling, B.:* Cross Polarization — Magic Angle Spinning NMR Studies of Carbohydrates and Aromatic Polymers. Vol. 66, pp. 1–22.
*Lipatov, Y. S.:* Relaxation and Viscoelastic Properties of Heterogeneous Polymeric Compositions. Vol. 22, pp. 1–59.
*Lipatov, Y. S.:* The Iso-Free-Volume State and Glass Transitions in Amorphous Polymers: New Development of the Theory. Vol. 26, pp. 63–104.
*Lipatova, T. E.:* Medical Polymer Adhesives. Vol. 79, pp. 65–93.
*Lohse, F., Zweifel, H.:* Photocrosslinking of Epoxy Resins. Vol. 78, pp. 59–80.
*Lustoň, J.* and *Vašš, F.:* Anionic Copolymerization of Cyclic Ethers with Cyclic Anhydrides. Vol. 56, pp. 91–133.

*Madec, J.-P.* and *Maréchal, E.:* Kinetics and Mechanisms of Polyesterifications. II. Reactions of Diacids with Diepoxides. Vol. 71, pp. 153–228.
*Mano, E. B.* and *Coutinho, F. M. B.:* Grafting on Polyamides. Vol. 19, pp. 97–116.
*Maréchal, E.* see Madec, J.-P. Vol. 71, pp. 153–228.
*Mark, J. E.:* The Use of Model Polymer Networks to Elucidate Molecular Aspects of Rubberlike Elasticity. Vol. 44, pp. 1–26.
*Mark, J. E.* see Queslel, J. P. Vol. 71, pp. 229–248.
*Maser, F., Bode, K., Pillai, V. N. R.* and *Mutter, M.:* Conformational Studies on Model Peptides.

Their Contribution to Synthetic, Structural and Functional Innovations on Proteins. Vol. 65, pp. 177–214.

*McGrath, J. E.* see Yorkgitis, E. M. Vol. 72, pp. 79–110.

*McIntyre, J. E.* see Dobb, M. G. Vol. 60/61, pp. 61–98.

*Meerwall v., E. D.:* Self-Diffusion in Polymer Systems. Measured with Field-Gradient Spin Echo NMR Methods, Vol. 54, pp. 1–29.

*Mengoli, G.:* Feasibility of Polymer Film Coating Through Electroinitiated Polymerization in Aqueous Medium. Vol. 33, pp. 1–31.

*Mertzel, E., Koenik, J. L.:* Application of FT-IR and NMR to Epoxy Resins. Vol. 75, pp. 73–112.

*Meyerhoff, G.:* Die viscosimetrische Molekulargewichtsbestimmung von Polymeren. Vol. 3, pp. 59–105.

*Millich, F.:* Rigid Rods and the Characterization of Polyisocyanides. Vol. 19, pp. 117–141.

*Möller, M.:* Cross Polarization — Magic Angle Sample Spinning NMR Studies. With Respect to the Rotational Isomeric States of Saturated Chain Molecules. Vol. 66, pp. 59–80.

*Morawetz, H.:* Specific Ion Binding by Polyelectrolytes. Vol. 1, pp. 1–34.

*Morgan, R. J.:* Structure-Property Relations of Epoxies Used as Composite Matrices. Vol. 72, pp. 1–44.

*Morin, B. P., Breusova, I. P.* and *Rogovin, Z. A.:* Structural and Chemical Modifications of Cellulose by Graft Copolymerization. Vol. 42, pp. 139–166.

*Mulvaney, J. E., Oversberger, C. C.* and *Schiller, A. M.:* Anionic Polymerization. Vol. 3, pp. 106–138.

*Nakase, Y., Kurijama, I.* and *Odajima, A.:* Analysis of the Fine Structure of Poly(Oxymethylene) Prepared by Radiation-Induced Polymerization in the Solid State. Vol. 65, pp. 79–134.

*Neuse, E.:* Aromatic Polybenzimidazoles. Syntheses, Properties, and Applications. Vol. 47, pp. 1–42.

*Nicolais, L.* see Apicella, A. Vol. 72, pp. 69–78.

*Nuyken, O., Weidner, R.:* Graft and Block Copolymers via Polymeric Azo Initiators. Vol. 73/74, pp. 145–200.

*Ober, Ch. K., Jin, J.-I.* and *Lenz, R. W.:* Liquid Crystal Polymers with Flexible Spacers in the Main Chain. Vol. 59, pp. 103–146.

*Okubo, T.* and *Ise, N.:* Synthetic Polyelectrolytes as Models of Nucleic Acids and Esterases. Vol. 25, pp. 135–181.

*Oleinik, E. F.:* Epoxy-Aromatic Amine Networks in the Glassy State Structure and Properties. Vol. 80, pp. 49–99.

*Osaki, K.:* Viscoelastic Properties of Dilute Polymer Solutions. Vol. 12, pp. 1–64.

*Oster, G.* and *Nishijima, Y.:* Fluorescence Methods in Polymer Science. Vol. 3, pp. 313–331.

*Otsu, T.* see Sato, T. Vol. 71, pp. 41–78.

*Overberger, C. G.* and *Moore, J. A.:* Ladder Polymers. Vol. 7, pp. 113–150.

*Packirisamy, S.* see Biswas, M. Vol. 70, pp. 71–118.

*Papkov, S. P.:* Liquid Crystalline Order in Solutions of Rigid-Chain Polymers. Vol. 59, pp. 75–102.

*Patat, F., Killmann, E.* und *Schiebener, C.:* Die Absorption von Makromolekülen aus Lösung. Vol. 3, pp. 332–393.

*Patterson, G. D.:* Photon Correlation Spectroscopy of Bulk Polymers. Vol. 48, pp. 125–159.

*Penczek, S., Kubisa, P.* and *Matyjaszewski, K.:* Cationic Ring-Opening Polymerization of Heterocyclic Monomers. Vol. 37, pp. 1–149.

*Penczek, S., Kubisa, P.* and *Matyjaszewski, K.:* Cationic Ring-Opening Polymerization; 2. Synthetic Applications. Vol. 68/69, pp. 1–298.

*Penczek, S.* see Klosinski, P.: Vol. 79, pp. 139–157.

*Peticolas, W. L.:* Inelastic Laser Light Scattering from Biological and Synthetic Polymers. Vol. 9, pp. 285–333.

*Petropoulos, J. H.:* Membranes with Non-Homogeneous Sorption Properties. Vol. 64, pp. 85–134.

*Pino, P.:* Optically Active Addition Polymers. Vol. 4, pp. 393–456.

*Pitha, J.:* Physiological Activities of Synthetic Analogs of Polynucleotides. Vol. 50, pp. 1–16.

*Platé, N. A.* and *Noak, O. V.:* A Theoretical Consideration of the Kinetics and Statistics of Reactions of Functional Groups of Macromolecules. Vol. 31, pp. 133–173.
*Platé, N. A., Valuev, L. I.:* Heparin-Containing Polymeric Materials. Vol. 79, pp. 95–138.
*Platé, N. A.* see Shibaev, V. P. Vol. 60/61, pp. 173–252.
*Plesch, P. H.:* The Propagation Rate-Constants in Cationic Polymerisations. Vol. 8, pp. 137–154.
*Porod, G.:* Anwendung und Ergebnisse der Röntgenkleinwinkelstreuung in festen Hochpolymeren. Vol. 2, pp. 363–400.
*Pospíšil, J.:* Transformations of Phenolic Antioxidants and the Role of Their Products in the Long-Term Properties of Polyolefins. Vol. 36, pp. 69–133.
*Postelnek, W., Coleman, L. E.,* and *Lovelace, A. M.:* Fluorine-Containing Polymers. I. Fluorinated Vinyl Polymers with Functional Groups, Condensation Polymers, and Styrene Polymers. Vol. 1, pp. 75–113.

*Queslel, J. P.* and *Mark, J. E.:* Molecular Interpretation of the Moduli of Elastomeric Polymer Networks of Know Structure. Vol. 65, pp. 135–176.
*Queslel, J. P.* and *Mark, J. E.:* Swelling Equilibrium Studies of Elastomeric Network Structures. Vol. 71, pp. 229–248.

*Rehage, G.* see Finkelmann, H. Vol. 60/61, pp. 99–172.
*Rempp, P. F.* and *Franta, E.:* Macromonomers: Synthesis, Characterization and Applications. Vol. 58, pp. 1–54.
*Rempp, P., Herz, J.* and *Borchard, W.:* Model Networks. Vol. 26, pp. 107–137.
*Richards, R. W.:* Small Angle Neutron Scattering from Block Copolymers. Vol. 71, pp. 1–40.
*Rigbi, Z.:* Reinforcement of Rubber by Carbon Black. Vol. 36, pp. 21–68.
*Rogovin, Z. A.* and *Gabrielyan, G. A.:* Chemical Modifications of Fibre Forming Polymers and Copolymers of Acrylonitrile. Vol. 25, pp. 97–134.
*Roha, M.:* Ionic Factors in Steric Control. Vol. 4, pp. 353–392.
*Roha, M.:* The Chemistry of Coordinate Polymerization of Dienes. Vol. 1, pp. 512–539.
*Rostami, S.* see Walsh, D. J. Vol. 70, pp. 119–170.
*Rozengerk, v. A.:* Kinetics, Thermodynamics and Mechanism of Reactions of Epoxy Oligomers with Amines. Vol. 75, pp. 113–166.

*Safford, G. J.* and *Naumann, A. W.:* Low Frequency Motions in Polymers as Measured by Neutron Inelastic Scattering. Vol. 5, pp. 1–27.
*Sato, T.* and *Otsu, T.:* Formation of Living Propagating Radicals in Microspheres and Their Use in the Synthesis of Block Copolymers. Vol. 71, pp. 41–78.
*Sauer, J. A.* and *Chen, C. C.:* Crazing and Fatigue Behavior in One and Two Phase Glassy Polymers. Vol. 52/53, pp. 169–224.
*Sawamoto, M.* see Higashimura, T. Vol. 62, pp. 49–94.
*Schmidt, R. G., Bell, J. P.:* Epoxy Adhesion to Metals. Vol. 75, pp. 33–72.
*Schuerch, C.:* The Chemical Synthesis and Properties of Polysaccharides of Biomedical Interest. Vol. 10, pp. 173–194.
*Schulz, R. C.* und *Kaiser, E.:* Synthese und Eigenschaften von optisch aktiven Polymeren. Vol. 4, pp. 236–315.
*Seanor, D. A.:* Charge Transfer in Polymers. Vol. 4, pp. 317–352.
*Semerak, S. N.* and *Frank, C. W.:* Photophysics of Excimer Formation in Aryl Vinyl Polymers, Vol. 54, pp. 31–85.
*Seidl, J., Malinský, J., Dušek, K.* und *Heitz, W.:* Makroporöse Styrol-Divinylbenzol-Copolymere und ihre Verwendung in der Chromatographie und zur Darstellung von Ionenaustauschern. Vol. 5, pp. 113–213.
*Semjonow, V.:* Schmelzviskositäten hochpolymerer Stoffe. Vol. 5, pp. 387–450.
*Semlyen, J. A.:* Ring-Chain Equilibria and the Conformations of Polymer Chains. Vol. 21, pp. 41–75.
*Sen, A.:* The Copolymerization of Carbon Monoxide with Olefins. Vol. 73/74, pp. 125–144.
*Senturia, S. D., Sheppard, N. F. Jr.:* Dielectric Analysis of Thermoset Cure. Vol. 80, pp. 1–47.

*Sharkey, W. H.:* Polymerizations Through the Carbon-Sulphur Double Bond. Vol. 17, pp. 73–103.
*Sheppard, N. F. Jr.* see Senturia, S. D.: Vol. 80, pp. 1–47.
*Shibaev, V. P.* and *Platé, N. A.:* Thermotropic Liquid-Crystalline Polymers with Mesogenic Side Groups. Vol. 60/61, pp. 173–252.
*Shimidzu, T.:* Cooperative Actions in the Nucleophile-Containing Polymers. Vol. 23, pp. 55–102.
*Shutov, F. A.:* Foamed Polymers Based on Reactive Oligomers, Vol. 39, pp. 1–64.
*Shutov, F. A.:* Foamed Polymers. Cellular Structure and Properties. Vol. 51, pp. 155–218.
*Shutov, F. A.:* Syntactic Polymer Foams. Vol. 73/74, pp. 63–124.
*Siesler, H. W.:* Rheo-Optical Fourier-Transform Infrared Spectroscopy: Vibrational Spectra and Mechanical Properties of Polymers. Vol. 65, pp. 1–78.
*Silvestri, G., Gambino, S.,* and *Filardo, G.:* Electrochemical Production of Initiators for Polymerization Processes. Vol. 38, pp. 27–54.
*Sixl, H.:* Spectroscopy of the Intermediate States of the Solid State Polymerization Reaction in Diacetylene Crystals. Vol. 63, pp. 49–90.
*Slichter, W. P.:* The Study of High Polymers by Nuclear Magnetic Resonance. Vol. 1, pp. 35–74.
*Small, P. A.:* Long-Chain Branching in Polymers. Vol. 18.
*Smets, G.:* Block and Graft Copolymers. Vol. 2, pp. 173–220.
*Smets, G.:* Photochromic Phenomena in the Solid Phase. Vol. 50, pp. 17–44.
*Sohma, J.* and *Sakaguchi, M.:* ESR Studies on Polymer Radicals Produced by Mechanical Destruction and Their Reactivity. Vol. 20, pp. 109–158.
*Solaro, R.* see Chiellini, E. Vol. 62, pp. 143–170.
*Sotobayashi, H.* und *Springer, J.:* Oligomere in verdünnten Lösungen. Vol. 6, pp. 473–548.
*Sperati, C. A.* and *Starkweather, Jr., H. W.:* Fluorine-Containing Polymers. II. Polytetrafluoroethylene. Vol. 2, pp. 465–495.
*Spiess, H. W.:* Deuteron NMR — A new Toolfor Studying Chain Mobility and Orientation in Polymers. Vol. 66, pp. 23–58.
*Sprung, M. M.:* Recent Progress in Silicone Chemistry. I. Hydrolysis of Reactive Silane Intermediates, Vol. 2, pp. 442–464.
*Stahl, E.* and *Brüderle, V.:* Polymer Analysis by Thermofractography. Vol. 30, pp. 1–88.
*Stannett, V. T., Koros, W. J., Paul, D. R., Lonsdale, H. K.,* and *Baker, R. W.:* Recent Advances in Membrane Science and Technology. Vol. 32, pp. 69–121.
*Staverman, A. J.:* Properties of Phantom Networks and Real Networks. Vol. 44, pp. 73–102.
*Stauffer, D., Coniglio, A.* and *Adam, M.:* Gelation and Critical Phenomena. Vol. 44, pp. 103–158.
*Stille, J. K.:* Diels-Alder Polymerization. Vol. 3, pp. 48–58.
*Stolka, M.* and *Pai, D.:* Polymers with Photoconductive Properties. Vol. 29, pp. 1–45.
*Stuhrmann, H.:* Resonance Scattering in Macromolecular Structure Research. Vol. 67, pp. 123–164.
*Subramanian, R. V.:* Electroinitiated Polymerization on Electrodes. Vol. 33, pp. 35–58.
*Sumitomo, H.* and *Hashimoto, K.:* Polyamides as Barrier Materials. Vol. 64, pp. 55–84.
*Sumitomo, H.* and *Okada, M.:* Ring-Opening Polymerization of Bicyclic Acetals, Oxalactone, and Oxalactam. Vol. 28, pp. 47–82.
*Szegö, L.:* Modified Polyethylene Terephthalate Fibers. Vol. 31, pp. 89–131.
*Szwarc, M.:* Termination of Anionic Polymerization. Vol. 2, pp. 275–306.
*Szwarc, M.:* The Kinetics and Mechanism of N-carboxy-α-amino-acid Anhydride (NCA) Polymerization to Poly-amino Acids. Vol. 4, pp. 1–65.
*Szwarc, M.:* Thermodynamics of Polymerization with Special Emphasis on Living Polymers. Vol. 4, pp. 457–495.
*Szwarc, M.:* Living Polymers and Mechanisms of Anionic Polymerization. Vol. 49, pp. 1–175.


*Takahashi, A.* and *Kawaguchi, M.:* The Structure of Macromolecules Adsorbed on Interfaces. Vol. 46, pp. 1–65.
*Takemoto, K.* and *Inaki, Y.:* Synthetic Nucleic Acid Analogs. Preparation and Interactions. Vol. 41, pp. 1–51.
*Tani, H.:* Stereospecific Polymerization of Aldehydes and Epoxides. Vol. 11, pp. 57–110.
*Tate, B. E.:* Polymerization of Itaconic Acid and Derivatives. Vol. 5, pp. 214–232.
*Tazuke, S.:* Photosensitized Charge Transfer Polymerization. Vol. 6, pp. 321–346.

*Teramoto, A.* and *Fujita, H.:* Conformation-dependent Properties of Synthetic Polypeptides in the Helix-Coil Transition Region. Vol. 18, pp. 65–149.
*Theocaris, P. S.:* The Mesophase and its Influence on the Mechanical Behvior of Composites. Vol. 66, pp. 149–188.
*Thomas, W. M.:* Mechanismus of Acrylonitrile Polymerization. Vol. 2, pp. 401–441.
*Tieke, B.:* Polymerization of Butadiene and Butadiyne (Diacetylene) Derivatives in Layer Structures. Vol. 71, pp. 79–152.
*Tobolsky, A. V.* and *DuPré, D. B.:* Macromolecular Relaxation in the Damped Torsional Oscillator and Statistical Segment Models. Vol. 6, pp. 103–127.
*Tosi, C.* and *Ciampelli, F.:* Applications of Infrared Spectroscopy to Ethylene-Propylene Copolymers. Vol. 12, pp. 87–130.
*Tosi, C.:* Sequence Distribution in Copolymers: Numerical Tables. Vol. 5, pp. 451–462.
*Tran, C.* see Yorkgitis, E. M. Vol. 72, pp. 79–110.
*Tsuchida, E.* and *Nishide, H.:* Polymer-Metal Complexes and Their Catalytic Activity. Vol. 24, pp. 1–87.
*Tsuji, K.:* ESR Study of Photodegradation of Polymers. Vol. 12, pp. 131–190.
*Tsvetkov, V.* and *Andreeva, L.:* Flow and Electric Birefringence in Rigid-Chain Polymer Solutions. Vol. 39, pp. 95–207.
*Tuzar, Z., Kratochvil, P.,* and *Bohdanecký, M.:* Dilute Solution Properties of Aliphatic Polyamides. Vol. 30, pp. 117–159.

*Uematsu, I.* and *Uematsu, Y.:* Polypeptide Liquid Crystals. Vol. 59, pp. 37–74.

*Valuev, L. I.* see Platé, N. A.: Vol. 79, pp. 95–138.
*Valvassori, A.* and *Sartori, G.:* Present Status of the Multicomponent Copolymerization Theory. Vol. 5, pp. 28–58.
*Vidal, A.* see Donnet, J. B. Vol. 76, pp. 103–128.
*Viovy, J. L.* and *Monnerie, L.:* Fluorescence Anisotropy Technique Using Synchrotron Radiation as a Powerful Means for Studying the Orientation Correlation Functions of Polymer Chains. Vol. 67, pp. 99–122.
*Voigt-Martin, I.:* Use of Transmission Electron Microscopy to Obtain Quantitative Information About Polymers. Vol. 67, pp. 195–218.
*Voorn, M. J.:* Phase Separation in Polymer Solutions. Vol. 1, pp. 192–233.

*Walsh, D. J., Rostami, S.:* The Miscibility of High Polymers: The Role of Specific Interactions. Vol. 70, pp. 119–170.
*Ward, I. M.:* Determination of Molecular Orientation by Spectroscopic Techniques. Vol. 66, pp. 81–116.
*Ward, I. M.:* The Preparation, Structure and Properties of Ultra-High Modulus Flexible Polymers. Vol. 70, pp. 1–70.
*Weidner, R.* see Nuyken, O.: Vol. 73/74, pp. 145–200.
*Werber, F. X.:* Polymerization of Olefins on Supported Catalysts. Vol. 1, pp. 180–191.
*Wichterle, O., Šebenda, J.,* and *Králiček, J.:* The Anionic Polymerization of Caprolactam. Vol. 2, pp. 578–595.
*Wilkes, G. L.:* The Measurement of Molecular Orientation in Polymeric Solids. Vol. 8, pp. 91–136.
*Wilkes, G. L.* see Yorkgitis, E. M. Vol. 72, pp. 79–110.
*Williams, G.:* Molecular Aspects of Multiple Dielectric Relaxation Processes in Solid Polymers. Vol. 33, pp. 59–92.
*Williams, J. G.:* Applications of Linear Fracture Mechanics. Vol. 27, pp. 67–120.
*Wöhrle, D.:* Polymere aus Nitrilen. Vol. 10, pp. 35–107.
*Wöhrle, D.:* Polymer Square Planar Metal Chelates for Science and Industry. Synthesis, Properties and Applications. Vol. 50, pp. 45–134.
*Wolf, B. A.:* Zur Thermodynamik der enthalpisch und der entropisch bedingten Entmischung von Polymerlösungen. Vol. 10, pp. 109–171.

*Woodward, A. E.* and *Sauer, J. A.:* The Dynamic Mechanical Properties of High Polymers at Low Temperatures. Vol. 1, pp. 114–158.

*Wunderlich, B.:* Crystallization During Polymerization. Vol. 5, pp. 568–619.

*Wunderlich, B.* and *Baur, H.:* Heat Capacities of Linear High Polymers. Vol. 7, pp. 151–368.

*Wunderlich, B.* and *Grebowicz, J.:* Thermotropic Mesophases and Mesophase Transitions of Linear, Flexible Macromolecules. Vol. 60/61, pp. 1–60.

*Wrasidlo, W.:* Thermal Analysis of Polymers. Vol. 13, pp. 1–99.


*Yamashita, Y.:* Random and Black Copolymers by Ring-Opening Polymerization. Vol. 28, pp. 1–46.

*Yamazaki, N.:* Electrolytically Initiated Polymerization. Vol. 6, pp. 377–400.

*Yamazaki, N.* and *Higashi, F.:* New Condensation Polymerizations by Means of Phosphorus Compounds. Vol. 38, pp. 1–25.

*Yokoyama, Y.* and *Hall, H. K.:* Ring-Opening Polymerization of Atom-Bridged and Bond-Bridged Bicyclic Ethers, Acetals and Orthoesters. Vol. 42, pp. 107–138.

*Yorkgitis, E. M., Eiss, N. S. Jr., Tran, C., Wilkes, G. L.* and *McGrath, J. E.:* Siloxane-Modified Epoxy Resins. Vol. 72, pp. 79–110.

*Yoshida, H.* and *Hayashi, K.:* Initiation Process of Radiation-induced Ionic Polymerization as Studied by Electron Spin Resonance. Vol. 6, pp. 401–420.

*Young, R. N., Quirk, R. P.* and *Fetters, L. J.:* Anionic Polymerizations of Non-Polar Monomers Involving Lithium. Vol. 56, pp. 1–90.

*Yuki, H.* and *Hatada, K.:* Stereospecific Polymerization of Alpha-Substituted Acrylic Acid Esters. Vol. 31, pp. 1–45.


*Zachmann, H. G.:* Das Kristallisations- und Schmelzverhalten hochpolymerer Stoffe. Vol. 3, pp. 581–687.

*Zaikov, G. E.* see Aseeva, R. M. Vol. 70, pp. 171–230.

*Zakharov, V. A., Bukatov, G. D.,* and *Yermakov, Y. I.:* On the Mechanism of Olifin Polymerization by Ziegler-Natta Catalysts. Vol. 51, pp. 61–100.

*Zambelli, A.* and *Tosi, C.:* Stereochemistry of Propylene Polymerization. Vol. 15, pp. 31–60.

*Zucchini, U.* and *Cecchin, G.:* Control of Molecular-Weight Distribution in Polyolefins Synthesized with Ziegler-Natta Catalytic Systems. Vol. 51, pp. 101–154.

*Zweifel, H.* see Lohse, F.: Vol. 78, pp. 59–80.

# Subject Index

Activation energy 68, 84, 147
— enthalpy 84
— volume 83
Admittance, definition 5, 6
—, linearity 4
—, measurement 4
—, — equipment 12, 13
—, time invariance 5
Aliphatic chain conformation 69
— chains, flexible 71
Amines, aliphatic 54
—, aromatic 52
—, tetra-reacted 88
Aminolysis 54
Anhydride-cured resin 187
Aniline 53
Anionic catalysts 176
Annealing 195
—, thermal 133
Aromatic anhydrides 188
Arrhenius equation 146
Autoassociation 65
Autoclave process 102

Beta loss peak 142
— -relaxation 141
— -transition 143
Boron trifluoride 128
Brittleness 75
Bubble growth 110

Carboxylic hydrazides 183
Catalysts 176
Cationic catalysts 176
Chain extensibility 77
Chemical crosslinking 82
— defects 56, 58, 80
Cleavage 95
Cohesion energy 66
Cohesive energy density 194
Cole-Cole diagram 17–19, 22, 23
Component ratio, initial 80

Composites, advanced 51, 75
—, matrix for 51
Computer simulations 57
Configurational mobility 62
Conformational bonds 70
— elasticity 76
Conformers 70
Consolidation theories 121
Conversion topological limit 55
CP/MAS NMR 158
Crankshaft 68
— motion 141
Craze 83
Critical conversion 86
Crosslink connectivity 59, 73, 84
— density 56, 80
— functionality 56
—, 3-linked 96
— mobility 74
—, spatial distribution 97
—, — inhomogeneities 59
—, structure 183
Crosslinking, degree of 177
— density 178
— reactions 135
Cure 52, 88
— contraction 90
—, isothermal 91
— time 91
Curing mechanisms with epoxy resins 174
— — with anhydride 176
— — with amines 174
— — with mercaptan 188
Cycloaliphatic acid anhydrides 188

Darcy's law 119, 120, 121
Debye analysis 163
— model 16–18
Deformation, crystal-like 85
— hardening 79
—, inhomogeneous 83
—, irreversible 85
— recovery 84

Density 92, 94
— fluctuation 94
— measurements 151
Diamine curing agents 181
Dielectric dissipation factor 7
— loss factor 16, 17, 22
— — —, definition 8
— — —, of curing epoxy 26
— — —, of DGEBA epoxy resin 25
— — tangent 7, 9, 23, 24
— — —, definition 8
— measurement, bibliography 40, 41
— — equipment 12, 13
— permittivity 17, 22
— —, definition 8
— —, of curing epoxy 26
— —, of DGEBA epoxy resin 25
— —, relative 8
— —, relaxed 16, 29—32
— —, unrelaxed 16
Diepoxide chain 76
Differential scanning calorimetry 130
Diffusion control 55
— conversion limit 72
Diffusion limit 56
Diffusional plasticity 85
Diffusivity 156
Diglycidyl ethers 52, 80
— — of resorcinol 53
— — of bisphenol-A 53
Diisocyanate 193
Diphenylmethane 193
Dipole loss peak 35
— moment 68
— —, average 69
— relaxation 67
— — functions 19, 20
— — time 16, 33, 34
— — — distribution 35, 36
Ductility 83, 133
Dynamic mechanical analysis 130, 140f.
— — properties 174, 178, 191, 193
— shear modulus 198
— storage modulus 140
— Young's modulus 178

Effective modulus 77
Elastically active chains 77
Elasticity modulus 77, 194
Electrode polarization 20—25, 27, 40
Electrodes, comb 9, 10
—, other 12
—, parallel plate 8, 9
Elongation 76
— at break 76
Embrittlement 82, 135
Engineering plastic 75

Enthalpy relaxation 144, 145
Epoxy-aromatic amine networks 49ff.
Epoxies, ester-type 55
Epoxy glasses 51, 90
— —, time-dependent behavior 127
— laminates 129
— matrices 127
— -matrix composite, embrittlement 135
— -moisture interactions 125, 131, 153
— ring 54
Equilibrium glassy state 151
— rubbery modulus 59
— shear modulus 76
Ether chain, flexible 97
— -type crosslinks 54
Expansivity, thermal 147
—, glassy state 149
—, rubbery-state 149

Flexibility 70
Flexural modulus 195
— strength 196
Fracture 50, 94
— initiation 96
Free induction decay (FID) 131
Free volume 60, 72, 94
Front factor 179
FTIR 144
Functional groups 174

Gelation 27, 28, 33, 38, 39
Glasses, crosslinked 51
Glass network 93
Glass transition, interval 91
— — temperature 49, 71, 73, 178
— — —, experimental 72
Glassy region 178
— state 49, 50, 51, 60, 194
Glycidyl groups 81
Guinier analysis 164

Hardness measurements 152
Heat capacity 61, 62, 90
— resistant polymers 72
Heterocurrent 67
Hole concentration 64
— formation, energy of 64
— theory 63
Hydrogen bonds in epoxy networks 65f.

2-Imino-oxazolidine rings 193
Impact strength 197
Inplane shear stress-strain tests 138
Internal rotation 70, 71
Ionic conductivity 15—17, 26, 37—40
Ion mobility 15, 16
Ions, migration 67

Isocyanates, curing 192
Isocyanurate ring 193

Laminate consolidation, resin flow 102, 103
Laminating process 102
Lattice crosslinks, defective 58
Limiting conversion 86
Local elongation 79
— mode relaxation 198
— plasticity 96
Long-term strength 95
Lubrication theory 120

Macrofracture 95
Maximum dipole loss, frequency 26, 32—34
Maxwell-Wagner effect 24, 25
Mechanical analysis, thermal 130, 147ff.
— —, dynamic 130, 140f.
— dispersion 143
Mechanical properties 78, 80, 92, 174
Mercaptan hardener 188
4,4'-Methylene bis(phenyl cyanamide) 193
Microdielectrometry 10
—, calibration 11, 14
—, instrumentation 14
—, sensor 10
Mobility of reacting groups 72
Modulus 138
— of elasticity 194
Moisture sorption kinetics 152
Molecular cycles 56
— mobility 61
— relaxation time 136
— volume 73
— weight between linkings 178
Monocycles 57, 58, 80
Monte-Carlo simulation 95

Network connectivity 54
— fragments, immobile 97
— topological nonuniformity 59
NMR 131
—, CP/MAS 158
—, solid state magic-angle-spinning 163
Non-equilibrium glassy state 126
Notch 96

Overloading of chemical bond 95
Oxazolidone 193

Packing density 60, 71, 81, 97
Partial heating procedure 68
Peak cleaning procedure 67
Permeability in laminate 103
$m$-Phenylenediamine 52, 53
Phenylglycidine ether 74
Phenylglycidyl ether 53

Physical aging 82, 125, 126
— —, free-volume concept 127
— —, thermoreversibility 129, 152
— —, toughness 134
— —, yield strength 134
Plastic deformation 83
— nuclei 84
— zone 96
Positron annihilation 166
Post-cure 50, 75
Polyamine 181
Polyanhydrides, linear 187
Polyisobutylene 63
Polymercaptans 189
Polymerization catalysts 176
Polymer-solvent interactions 156
Polystyrene 63

Quenching 86, 195

Raoult's law 107
Relaxation 89
— of small structural units 198
—, secondary 141
— time distribution 75
Residual thermal stress 136
Resin flow during laminate consolidation 102, 103
— — model 119, 121
— — —, three-dimensional 122
— pressures 109, 120
— — gradient 122
— — profiles 120, 121, 122
— transport 119
— viscosity 119
Resistivity 16
Rigidity 77
— of main chain 180
Ring-current effects 158
Rubbery region 178
Rubbery state 50, 75

Saturation polarization 68, 69
SAXS 131, 162, 163, 166
Shear band 83
— defects 85, 94
— moduli 81
— modulus, ten-second 195
Simha-Boyer factor 64
Small-angle X-ray scattering (SAXS) 131, 162, 163, 166
Softening 88
Solubility parameter 194
Sorption, dual mode 155
— kinetics 153
Specific heat 55
Strain rate 79

Strength of interaction 180
Stress concentration coefficient 95
— relaxation 129
— — analysis 138
— -strain analysis 132
— -— curve 78
Structural defects 57
Structure of cured resin 174
Sub-$T_g$ annealing 82, 127, 137
Supercooled liquids 126
Swelling efficiency 155, 157

Tensile or bending strengths 195, 196
Thermal contraction 91
— expansion 61, 62
— mechanical analysis 130, 140f.
— treatment 50
Thermally stimulated discharge (TSD) 67
Thermodynamic functions 82
Topological limit 86
— structure 78
Topology 49, 56
Toughness 136
Transition, cooperativity 93
— region 178
α-Transition 74
—, shape 93
β-Transition 198
TSD current 93
TTT-diagram 87
Two-phase system 192

Ultimate tensile strength 136
Unoccupied volume 61
Unreacted epoxy end 69
— groups 87, 96

Van-der-Waals interactions 66
Viscosity 15, 16, 23, 33, 34, 36, 40
Vitrification 27, 28, 33, 35, 86
— during cure 55
— point 90

Void dissolution 116
—, final diameter 117
— formation 104, 105
—, gas density 112
— growth 122
— — calculations 111, 112
— —, diffusion-controlled 109
— — driving force 110
— —, equilibrium 109
— — model 110
— —, model predictions 114ff.
— — —, pure water vapor 109
— — /stability, time-dependent 121
— model framework 104
— nucleation 109, 111
— ratio 120
— size, final 116, 117, 118, 122
— —, initial 122
— stability at equilibrium 106
— — map 108, 109, 119
— —, phase map 121
— stabilization 106
— transport 119
—, vapor pressure 112
Voids 102ff.
—, equilibrium stability 105
—, heterogeneous nucleation 106
—, homogeneous nucleation 105
—, kinetics and viscosity models 104
—, mechanical entrapment 105
—, nucleation processes 105
—, pressure-temperature stability map 122
—, resin water concentration 112
Volume, specific 90

Williams-Landel-Ferry equation 27, 33—39

Yield strength 196
— stress 79, 81, 92
Young modulus 60, 66, 81, 92, 93
— —, dynamic 87, 91

# Epoxy Resins and Composites I

Editor: K. Dušek

1985. 79 figures, 16 tables. XIII, 167 pages. (Advances in Polymer Science, Volume 72). ISBN 3-540-15546-5
Distribution rights for all socialist countries: Akademie-Verlag, Berlin

**Contents/Information:** The latest volume in Advances in Polymer Science contains five major studies:
*R. J. Morgan:* **Structure-Property Relations of Epoxies Used as Composite Matrices**
The first essay presents the structure-deformation/failure process-mechanical property relations of epoxies used as matrices in high performance fibrous composites.
*A. J. Kinloch:* **Mechanics and Mechanisms of Fracture of Thermosetting Epoxy Polymers**
The next section discusses the fracture of epoxy polymers, concentrating on the use of the continuum fracture mechanics approach for elucidating the micromechanisms of crack growth and identifying pertinent failure criterion.
*A. Apicella, L. Nicolais:* **Effect of Water on the Properties of Epoxy Matrix and Composite**
In this chapter the complex sorption behavior of the water in amineepoxy thermosets is discussed and related to depression of the mechanical properties.
*E. M. Yorkgitis, C. Tran, N. S. Eiss, J. E. McGrath, G. L. Wilkes:* **Siloxane-Modified Epoxy Resins**
Here is a full description of epoxy resins chemically modified with functionally terminated polydimethylsiloxane, polydimethyl-co-methyltrifluoropropyl siloxane, and polydimethyl-co-diphenyl siloxane oligomers in terms of their synthesis, morphology, solid-state properties, and friction and wear properties.
*J. M. Barton:* **The Application of Differential Scanning Calorimetry (DSC) to the Study of Epoxy Resins Curing Reactions**
The last chapter reviews the use of differential scanning calorimetry as a method of monitoring and investigating the kinetics of epoxy resin curing reactions.

Springer-Verlag
Berlin Heidelberg
New York Tokyo

# Epoxy Resins and Composites II

Editor: **K. Dušek**

1986. 78 figures, 22 tables. XIII, 180 pages. (Advances in Polymer Science, Volume 75). ISBN 3-540-15825-1
Distribution rights for all socialist countries: Akademie-Verlag, Berlin

**Contents:** *L. T. Drzal:* **The Interface in Epoxy Composites.** – *R. G. Schmidt, J. P. Bell:* **Epoxy Adhesion to Metals.** – *E. Mertzel, J. L. Koenig:* **Application of FT-IR and NMR to Epoxy Resins.** – *B. A. Rozenberg:* **Kinetics, Thermodynamics and Mechanism of Reactions of Epoxy Oligomers with Amines.**

# Epoxy Resins and Composites III

Editor: **K. Dušek**

1986. 77 figures, 9 tables. Approx. 170 pages. (Advances in Polymer Science, Volume 78). ISBN 3-540-15936-3
Distribution rights for all socialist countries: Akademie-Verlag, Berlin

This volume is a further publication covering the intense development of applications of epoxy resins in traditional and newly developing areas. It consists of four contributions: *K. Dušek:* **Network Formation in Curing of Epoxy Resins.** K. Dušek analyses the basis of crosslinking theories and their applications to curing of epoxy resins. – *F. Lohse, H. Zweifel:* **Photocrosslinking of Epoxy Resins.** They review structures and reactivity of photoinitiators for epoxies. – *M. T. Aronhime, J. K. Gillham:* **The Time-Temperature Transformation (TTT) Cure Diagram of Thermosetting Polymeric Systems.** They deal with two recent models describing the cure process. – *D. J. LeMay, F. N. Kelley:* **Structure and Ultimate Properties of Epoxy Resins.** – They describe common epoxy thermosets which are glassy at ambient temperatures and characterized by a densely crosslinked microstructure.

Springer-Verlag
Berlin Heidelberg
New York Tokyo

RETURN TO